EXISTENCE

Valentin Matcas, M.Ed.

DEDICATION

I dedicate this book to everyone eager to learn and develop
continuously throughout life.

CONTENTS

1   Everything That Exists   1

2   Objective Realities   37

3   Intelligences, Specialized Existence, and Private Worlds   63

4   Life in all Existential Details   96

1 EVERYTHING THAT EXISTS

It is easy to distinguish what it is from what it is not, just by telling them apart. Since this is the simplistic, empirical approach while studying existence. Yet what exactly is existence itself? How or where does existence exist? Is existence limited to the human existence, human cognition, or the humankind? Yes, because all living beings are defined by existence, just because the term 'being' implies to be, to exist. However, the term 'real' implies the same. Because the concept 'existence' has a multitude of terms defining it, and now we have to consider them while studying existence in all details. And as you notice, we are not looking for a superficial, empirical study, because we already have one, since we already know that everything that exists exists and everything that does not exist does not exist. We want more, we want a comprehensive study of existence, from all perspectives: existential, rational, living, interconnected, social, objective, analytical, created, consensual, cognitive, natural, mathematical, consensual, developmental, scientific, fiat, empirical, subjective, philosophical, ideological, highjective, accurate, algorithmic, legal, and correspondent. Only that, from all these distinct perspectives of existence, we end up with a distinct type of existence, as we have to consider these.

But what exactly is there escaping the senses of perception of all living beings in this world? The nonexistent, certainly. And what exactly exists and takes place beyond existence itself? Again, it is the nonexistent. Yet this is the case only for us and from our perspective, since existence is relative to any observer. You have one existence defining you, and it is your own existence. And in this manner, you may exist even on your own, and you still exist, for yourself. Yet for others, if they never encounter and never interact with you directly and implicitly, then you do not exist for them, you are simply part of the inexistent according to them, and so are they for you, from your own perspective. Because there might be zillions of worlds and realities similar to ours, but if they never interact with us, then they never exist for us, while we never exist for them. And this is the case for all realities, because nothing exists objectively at the exterior of any reality, not even the concept of 'exterior' itself.

We already notice how existence is capable to define everything as existing or not, in this simple Boolean manner. And this is the case because existence stands towards the base of our world, defining everything within to exist, to be real, or simply to be, since it is the same concept. However, when you try to define existence itself, you cannot, since there is not much below it to form a meaningful base of knowledge defining it, but only the absent, the missing, the unreal, or the nonexistent, distinguishing existence trivially, by contrast. This is your only mean to define accurately existence now, as being different from the nonexistent, and this is why you end up with a Boolean depiction for existence. To be or not to be. But never to be and not to be. While existence defines everything else in a similar Boolean manner: you either exist or you do not. You either interconnect with others being part of their existence, or not. You are either alive, or not. You are either fulfilling throughout life, or not.

Throughout this book, we study existence entirely, through accurate facts, from all perspectives and in all circumstances, in order to understand it accurately. Furthermore, we use existence to define, understand, and explain everything related to humans and human life. And since existence stands at the base of everything that exists, this accounts for everything.

Yet let us see what this world knows about existence. Because now, after ages of human civilization, human development, and therefore human knowledge, this is what it is known about existence, that what exists exists and what does not exist does not exist. Or 'I think, therefore I am.' Or 'to be or not to be,' defining existence and everything else as already seen, in this specific Boolean manner.

Yet there is more to consider, since we have two choices now at the beginning of our model of existence. One is Boolean as already seen, having only the existent and the nonexistent in this world and nothing else in between, as in to be or not to be but nothing else, and the other one considering everything else in between, an entire range of probability and agreement defining consensually and even statistically an entire assumed or speculative existence, bypassing the accurate one. With the entire consensual existence part of the entire current human society, with quantum mechanics included. Yet quantum mechanics is based on advanced mathematics and many times on enforced assumptions and observations, which are consensual. And in this manner, quantum mechanics contorts the accurate existence into the probabilistic existence blended with the consensual existence, while accurate existence itself is Boolean, having only two states, the existent and the nonexistent, with nothing else. To be or not to be, and this is accuracy.

And it is meaningful to know the actual case defining existence, since we cannot have both the accurate and the consensually accurate, being mutually disjunctive. You cannot decide now consensually among your ideology or jurisdiction that the sky is blue, since this is already the case. However, you can always decide consensually within your ideology or jurisdiction that the sky is yellow, and this becomes the consensual, ideological, juridical truth, that the sky is yellow. The blue color of the sky exists accurately, while the yellow color of the sky exists only consensually, ideologically, scientifically, or judicially, however you have managed to instate it, because this is how it exists now.

The sky is blue, and you do not have to agree upon it since the color blue of the sky exists accurately. But does the sky itself exists? Yes, since you can see it in the sky, it is there. Yet when you study the existence of the sky closely, you find it of an apparent type of existence, or empirical one, since you cannot go there to see it closely, objectively yourself. However, if you want a tangible existence, or an objective one, then you have to study it in all details. To notice how the optical light is dispersed in the blue color by the ozone molecules of the atmosphere. This is the accurate existence of the color blue of the sky, but not of the sky itself. Because the sky that you see is actually the entire space or universe out there beyond the orbit of Earth.

And now, does the universe that you see in the sky exist? Yes, since this is what the current science states. But does it exist accurately in this manner, or consensually, or only apparently? You cannot tell from the surface of Earth. Even more, the current science is consensual in nature, and you have to consider this, since in this manner, it confuses the consensual existence with the accurate existence, confusing you. And the more confused you are, the more controllable you become.

Yes, you may see the universe in the sky at night and during the day, with the sun, the moon, all stars, and all galaxies included, since you can see everything very clearly even by using very powerful telescopes. Yet this is only an empiric or apparent type of existence, but not necessarily the accurate existence. Because as long as you are not capable to perceive or distinguish the accurate existence firsthand, intelligently, and objectively, what is left is not the nonexistent, but it might be the consensual existent, or the apparent, empiric existent, the fiat existent, the algorithmic existent, or the ideological existent.

The current science and the entire current society trick you in every manner to confuse all types of existence in order to control you better, and this is why you always have to reason on your own. Or not, since many people love it under control, most of the humankind. But now, through their own composite consensual existence, they force you into anything they agree. And this is the first consensual developmental level.

While the current science is consensual entirely, based on agreements, which may be accurate or not, since science is based on an entire scientific consensus but not directly on the accurate existence. Because if science falsifies space records, then this is only the trickery or the fiat existence, but not the accurate existence. And now, if you do not have a firsthand, tangible, objective perception and therefore comprehension of the universe beyond the lower orbit of Earth, you cannot be certain if anything beyond the lower orbit of Earth exists. Space records always help, but if even these space records are inconclusive, you still cannot tell if the universe exists accurately beyond the orbit of Earth.

Did the big bang actually exists? How can you ever tell accurately, if you could not be there to witness it firsthand? Yes, there is a theory in this world associated to the big bang itself, the big bang theory, currently promoted by science consensually, yet the word theory means speculation and therefore not accuracy. Which means that the current science states clearly that the big bang took place in theory at the origin of the universe, and therefore not accurately. The big bang did not exist accurately but only theoretically or speculatively, theory that is accepted currently in a consensual manner by the entire scientific consensus of the current science.

As you notice, it is very important to know everything about existence, since existence stands at the base of this world, while standing at the base of your cognition, defining truth and accuracy entirely. More precisely, you place existence at the base of this world, in a conceptual manner. Because existence itself is a concept of another concept, and it is important to know. It is similar with the spacetime continuum, since the spacetime continuum, along with all systems of coordinate, do not exist accurately or objectively in this world or in the wider world, but only conceptually. And even so, existence remains very useful throughout cognition, similar to the spacetime continuum and all coordinate systems. Did you know that throughout history, scientists including Einstein believed in ether, or aether? Ether is an actual objective existence, or continuum, or grid, or coordinate system found everywhere in the universe, or this is what they claimed.

But no, existence itself is only conceptually present in this world, as a system of reference informing us of the existent and the nonexistent in this world. To be very precise now, you have a specific, very important cursor in your conscious mind, stating precisely with each concept, conception, and line of reasoning from your cognition, that they are real or not, that they exist or not, in your mind or in this world.

And this is how you can distinguish with ease the real, the erroneous, the imagined, the agreed to be real, the illusive, deceptive, enforced, legally instated, ideological, probabilistic, believed, and stereotypical, from the real, accurate truth, or accurate existence. While this extra existential cursor is very useful to maintain everything that you know and use throughout reasoning successful. Since this is how you may daydream with ease and in full confidence that your reveries are continuously distinguished from reality, and you are not schizophrenic, megalomaniac, or in a coma. However, if you are megalomaniac, schizophrenic, or in a coma, good luck to you, because now you use your illusive mind to check if your illusive mind is accurate, always stating that your mind is accurate. Which means that you are just fine, nothing to worry about.

But what if you have only lost your keys, and this has nothing to do with tyrants, the big bang, schizophrenia, and the blue color of the sky? Where exactly are your keys? Since your keys still exist now, somewhere, so is our mental model of existence that we form in this book capable enough to find your keys? As stated in the title of the book, this is the mental model of existence, and not of your keys. So we wish you good luck finding your keys. However, let us see what we can do.

Because, you are in your driveway now, there are no keys in your pockets, it is already late, you try to remember very well where you last had your keys, and you realize that your keys must be either in the living room, or in the bedroom. Where exactly are your keys? Your keys exist either in one room or the other, since they cannot be in two places simultaneously, while they cannot be in neither room either. Therefore, we have the Boolean 'to be or not to be' circumstance, and this seems to be the case for everything that exists or not, under all circumstances. Because this is the normal, accurate existence, just because accuracy is unique, making existence unique. Everything either exists or it does not. And it is the same with your keys. Only that physics itself had decided otherwise lately.

To be more precise, physics is the only discipline studying our world, this world, with everything objective part of it, along with all lines of causality holding these. Which means that physics itself must study everything that exists accurately objectively in this world. The classical physics does so, only that the latter addition to physics, the quantum physics, contorts classical physics, managing to contort the entire objective, accurate existence in this manner. But since the entire scientific consensus validates quantum physics, it must be consensually true, but not necessarily accurately true. While as you notice, you are better off looking for your keys on your own, because it might take us some time before finding them. And if we find them only consensually, they might not be there accurately.

Because quantum mechanics comes now to inform us that, since now you are more certain that the keys are in the living room, since you lose your keys in your living room more often than in the bedroom, giving us even their seventy-three percent probability of existence in the living room, now we know exactly where the keys are, in the living room. With a probability of seventy-three percent. And since everyone agrees, there is where your keys are, in the living room. And this is the probabilistic type of existence, but not the accurate existence.

However, quantum mechanics goes further, to inform us that your keys actually exist in both rooms simultaneously. More precisely, seventy-three percent of your keys exist in the living room, and twenty-seven percent of your keys exist in the bedroom. This is called quantum existence. It is not the normal accurate existence, even though quantum mechanics manages to confuse probabilistic existence with quantum existence and with accurate existence whenever there is no observer present to determine directly the accurate existence. Which is erroneous.

Because with an observer present, then yes, this is firsthand, objective, accurate existence. And do not trust too much this firsthand observer either, since there are many types of deceptive existence everywhere in this world. And if you place it in your cognition throughout your learning and entire experience in this world, you end up with an erroneous replica of the world in your mind, you end up judging everything through it, and this is how your reasoning and entire behavior in this world are erroneous, invalid, irrelevant, or diverted. Because if you cannot know your own world accurately, you cannot know your own place and meaning in this world, making your entire existence meaningless and irrelevant in this world, or even harmful.

What if you find one day a confused man on the street, actually suffering of total amnesia? What do you do? You simply inform him casually that his entire meaning in life and in this world is to clean up your house and cook all your meals. Which might seem inconsistent, but when you study this world, no one knows anything about meaning and fulfillment, and so he will serve you diligently from now on. But then when you study this world closely, you find all these tyrants throughout all upper social layers, profiting from a comprehensive exploitation in this world, always acting casually and innocently, since they cannot even know why this world serves them so diligently. Yes, the big bang existed accurately billions of years in the past, the tyrants inform you, forming this entire world spontaneously and randomly, and therefore you do not actually have a human meaning in this world, but you must serve them diligently. And so you do.

And if this does not seem too significant, imagine the two basketball teams starting to play the game of basketball without actually knowing the meaning or reason why they are there. What if they start playing volleyball instead, with the entire world watching them in astonishment? But what if the entire world does not know the meaning of everything in life and in this world? Are we not back to our example with schizophrenia and coma now, and with the tyrants too? And the keys? Where are the keys?

Because there is another type of existence present in quantum mechanics and in the entire current science, which is the consensual existence. Which again, is not the accurate existence, even if the current science and the entire Consensual Matrix seek to confuse the two, the consensual existence with the accurate existence.

While this is the 'Schrodinger's cat' assumption, that the keys exist probabilistically in two places at once, in an accurate manner, and in a consensual manner, simultaneously. Which is impossible, since probabilistic existence, consensual existence, and accurate existence are three distinct types of existence and cannot manifest simultaneously. Yet quantum mechanics does not use this probabilistic existential theory for macroscopic objects as keys, but only for elementary particles, with mathematics proving it right, probabilistically. Can existence be probabilistic, or only Boolean? Not simultaneously. The answer is obviously Boolean, as in the 'to be or not to be' case, or as in the 'what exists exists and what does not exist does not exist.' Which is the basic accurate existence.

But why exactly having multiple types of existence in this world, confusing this world, while existence should be only accurate, because everything in this world should be only accurate? This is the case not only with existence, but with everything in this world, since everything should be only accurate. However, throughout this book, we have to study all types of existence, naturally present in Life and in the wider world, along with everything invented and agreed by living beings, helpful or not to Life and the wider world. As the accurate existence, consensual existence, probabilistic existence, living existence called Existence, algorithmic existence, conceptual existence, manifested existence, consensual existence, correspondent existence, created existence, fiat existence, ideological existence, legal existence, and referential existence.

Yet let us keep all initial assumptions, just because there is more to this world than what we are made to believe. Living beings are capable to determine directly, bend directly, or manifest directly everything they choose, or everything they can, through higher abilities. These higher abilities are not allowed here in this world, or they are only a myth, a belief, or an illusion, yet if they have manifested at least once, it means that they are possible. Therefore, you can always manifest directly your keys wherever you want, yet even in this case, we notice an accurate model of existence, even when existence itself can be influenced by living beings directly, if this is ever possible.

And now, these few statements about existence, that what exists exists and what does no does not, are repeated throughout books and elaborated studies, as variations on the same theme, with the entire world feeling very comfortable with the level of knowledge in this specific subject. Yet go to places as Area 51, to see there what really exists. Because someone is taken for a ride, the entire world. Can it be related to these higher abilities? Is science fiction actually real?

What could be there more to know about existence, and why should anyone ever bother? Well, first, you never bother with anything important or not throughout life, unless your own needs for knowledge, learning, and development demand so. Learning is an intrinsic process and not extrinsic, as it follows curiosity. Learning may become an extrinsic process, many times throughout life, and this is called consensual education, mostly indoctrination. And this is how you are determined throughout life to learn and know more, both through curiosity and consensual education. Humans and all living beings follow this specific intrinsic need for knowledge felt as curiosity, everywhere and in every manner, only to be able to learn and develop, many times even by risking their lives or by having to leave their loved ones behind.

As an example, people walked to the poles or wandered throughout caves deeply underground only to fulfill their need for learning, which is humanity's need for learning and knowledge. Because as we always notice, life takes place not only at the level of the organism, but at the inner level of all intelligences of the human mind, at the level of all higher beings above, and at the level of the entire human society. Because human beings do not exist only for themselves but for everybody else, while learning, teaching, and providing for everybody else, including their loved ones. And many times, they do so even by sacrifice, since it always matters.

Humans exist for others and not only for themselves, but could it be that the human existence takes place not only in the human organism, not only in the human cognition, but in the human society and in all the higher worlds just as well? Can human beings be more than human, and more than individual beings? And it certainly helped if science knew more about existence, Life, Intelligence, humanity, Interconnectivity, human mind, and human needs.

This is not a simple model, since we have many topics to consider. And it is not exactly a matter of methodically recording everything that exists in this world in some database only to be able to know more, but many times, this is done as a matter of being able to distinguish between beliefs and accurate facts. Because, from today's perspective, you may assume that before the poles have been explored, there was this empty knowledge among humanity about what exists at the poles. No one simply knew what it was, but everybody was burning with curiosity. One day, people simply put together an expedition to reach the poles and see what was going on there, to make some drawings and take some notes while they were there, and so it happened. Yet life is more complex, because you never have to deal with the known and the unknown alone throughout life and throughout civilization, you never have to deal with the existent and with the inexistent, but with beliefs and accurate facts. You have to distinguish between beliefs and accurate facts, if you are capable enough to acknowledge these. And many times, this pushes you to learn and develop throughout life, your need to be able to distinguish between right and wrong continuously, and be able in this manner to keep your freedom. How heroic. Because while accurate facts define the existent alone while pointing out the inexistent, beliefs will define consensually both the existent and the inexistent alike, many times in an erroneous manner, for many reasons. Since this is the ideological, juridical, consensual existence. Do not underestimate it, since it can enslave most of the wider world, in one very strong consensus, the Consensual Matrix.

Because there was no empty knowledge about the poles before they had been explored, but the entire world was full of beliefs then, overfilling this blank ignorance about the poles. While these beliefs fueled dogma and debates in this world, not curiosity, as it reinforced the ignorance in this manner, but not the true values. Since the inexistent never comes in form of a blank concept ready to be explored and filled up in this manner with knowledge, but the inexistent is already filled up with beliefs, aberrations, and bad intentions of all kind, long before it is studied, explored, and understood. While only through learning, exploration, research, and development, you are ever capable to set things straight, to define what exists from what does not exist, through valid, genuine facts, and not through beliefs of any kind, including the strong sets of beliefs called ideologies, since this world is full of these.

And this is exactly the problem, because as long as you are not capable to distinguish between the existent and the inexistent on your own, by knowing well what existence is, you are doomed to fall into beliefs, stereotypes, and superstitions, whatever anyone wants you to believe and do. Because those around and the entire society will draw you into their dogma, and through it, they will use you throughout life for all purposes, and in every manner. It is called exploitation. Because the inexistent leads to ignorance, ignorance leads to beliefs, beliefs lead to ideologies and dogma, while dogma leads to servitude. Beliefs seem to lead you to servitude under these ideologies, yet beliefs actually lead you to servitude under those controlling these ideologies. This is how you lose your freedom throughout life, and it happens often, making possible all these tyrants found throughout the upper social levels both in the West and in the East. And it always starts with you being incapable to define what it is from what it is not, what exists from what it does not, and respectively, what it is accurate from what it is erroneous and irrelevant.

Because you always have to be capable to define existence in the first place. And this is why science states only that what exists exists and what does not exist does not, because this ignorance remains at the core of this outstanding social control spanning this world.

What is it there to know more than what you already do? What can it be there to exist beyond inexistence? Because by not being capable to understand and explain existence itself, how are you able to understand life in general and life on Earth in particular, life including you, your meaning, and your entire existence? And when people come back from the death with all kind of stories, how are you capable to understand these, if you cannot comprehend existence itself? And now you call these stories? What are dreams and astral projections when you cannot define existence itself? None of these stories exists, just because you can never bring with you plausible proof when you come back from the death. But where exactly have you been? How exactly do these places exist, if these are places, and how exactly do you exist while you are there?

Religion, brotherhoods, and spirituality mention the existence of higher, denser places, along with your own life there, after this one. While religion, brotherhoods, and spirituality are based on beliefs. You can never understand anything rigorously through beliefs, since beliefs are not facts. Beliefs do not even employ reasoning throughout cognition, but only simple memorization whenever you are demanded to do so throughout your servitude.

What can you do? Use only accurate facts throughout cognition, in the family, and in society. The only social domain using accurate facts is science. But science has the monopoly over all facts in this world in a totalitarian manner, while no one ever seems to notice or care. Even more, science remains ignorant and silent in everything concerning existence, and again, no one cares. Science is also full of beliefs, scientific beliefs, despite of what it may claim, with most of its consensual scientific beliefs being erroneous and misleading. While it also has a scientific consensus.

Therefore, if you ever want to understand anything in this world in an accurate manner, you have to study it yourself. While this is nothing new, since you have to do everything yourself throughout society, if you ever want to maintain your freedom and existence, since education dumbs you down, medicine kills you, science fools you while offending your intelligence, finance bankrupts you, and food industry poisons you.

How can anything exist after it ceases to exist? How can anything exist within other worlds independently of the existence that takes place here in this world? Is existence related directly to other worlds, or are worlds and realities related and determined by existence in general and by their own existence in particular? Because if existence itself influences everything around including you, life, and this world, then by ignoring and neglecting it, you miss understanding everything else, or you understand it erroneously, through beliefs. And this is the case just because existence stands at the base of everything.

To give an example, there are three concepts in physics that cannot be defined and explained independently today, but only one in rapport to another. These three concepts are of this world, and they are time, space, and matter or radiation.

Similarly, throughout this book series, I found three concepts of the wider world relating one to another. These three concepts are comprehensive and they do not limit to physics or to this world. These three concepts are the physical body, life, and intelligence. If you understand well these three concepts, you are able to understand and define everything in the wider world, including lifelines of existence making life eternal, along with higher and lower worlds, along with living beings, astral projection, time, space, and energy. And this is the case just because life, physical body, and intelligence are supreme concepts, and even more, because they are distinct perspectives of a oneness. These are distinct perspectives of you as a oneness, and they define you entirely. And at a supreme scale, these are supreme perspectives of the One himself. These concepts or perspectives are defined and determined independently by their own existence, while they are defined and determined as a whole by the existence of the one.

I had to use the word 'one' in order to define you in your entire lifeline of existence, since your life, body, and intelligence are more than connected directly, all three being correspondences of yourself. This means that you cannot be either the body, alive, or intelligent, if you are not these three simultaneously as one. You are not alive but you are more, you are not the physical body and alive because you are more, and you are not the physical body, alive, and intelligent because you are these three simultaneously, as one, you are one. You are one, because the only word I could use to define you as being the physical body, alive, and intelligent simultaneously, was 'one.' Since spirituality uses the word 'one' to define the One.

The One is everything, and this includes his Life, his Universal Mind, and everything that exists. While you are one in a very similar manner, at your own level. And now, through your own oneness, you are your life, your intelligence, and your physical body, simultaneously, as one, in one interconnectivity, and in your own one private reality. And this is the case because existence is correspondent for all living beings. And since this is the existence defining Life, Intelligence, and the wider world, as One, it does so in a correspondent manner. While I define it as Existence, the Existence of Life and the rest of her correspondences, Intelligence, Interconnectivity, and the wider world.

Therefore, you are not only objective, you are not only alive, and you are not only intelligent, but you are these as one. The reason why you cannot see yourself as one but only as being objective, alive, and intelligent, is because you lack a comprehensive perception and understanding of yourself, and therefore you are forced now to understand yourself piece by piece, through your three supreme perspectives separately: life, intelligence, and objective or physical body. However, you will always perceive yourself as one every time you interact with higher realities, through your own higher self, through astral projection, or through direct access to higher knowledge. And when you do so, you have to understand existence entirely and correspondently, since all ones have their own existence.

What is correspondence exactly? Everything is correspondent from one reality to another through you. Throughout life, you learn and experience everything in a normal, natural manner, placing everything that you perceive, understand, learn, and memorize in your mind, in your conscious intelligent mind, throughout the cortex. But you do not fill up entire databases or entire library of books there in your conscious intelligent mind, but you place all your memories, feelings, and understandings there exactly as you find them in this world, making in this manner a replica of this world in your mind. This replica of the world from your mind is a genuine reality, with all your memories and understandings as genuine concepts or conceptions, which are normal living intelligences living there in your inner replica of the world. We notice now a direct correspondence between all concepts, conceptions, memories, and understandings with their real counterparts from this world, along with a direct correspondence between your inner replica of the world and this world. Furthermore, the inner replica of yourself from this world, the physical body, is now an inner self in your inner replica of the world, consisting of everything that you can perceive, know, expect, and understand about yourself in this world.

You might always assume that you know everything about yourself, this world, and everything in this world, managing an entire accurate inner replica of the world in this manner. Which is never true, because you can never transfer anything from one reality to another, but only copied or correspondent information. But with everyone in this world assuming that they know everything accurately, it fuels all disagreements in this world, and now this is the world.

As a reference, the human knowledge is about one percent accurate, with the rest consensual, believed, erroneous, intoxicated, ignorant, altered, damaged, convicted, diverted, untrue, inexistent, irrelevant, undeveloped, stereotypical, enforced, ideological, supposed, theoretical, and juridical. One percent accuracy.

And this is only the difference between this world and the inner replica of the world, which is the mind. But you are not only mind and body, but mind, body, and soul, as one, and more, since you have more selves on your own lifeline of existence, as this is always correspondent. This world is made in the correspondent image of the higher world, where the souls live. While everything in the higher world is correspondent in this world, as the Creator of this world and the souls coming here made it possible. With all souls correspondent to the living human beings, as the living human beings are correspondent to their own inner selves.

Because you have your own existence defining you in this world and in your specific environment, among everybody that you interconnect with. While your entire environment and reality define you through their own existence just as well, and if they do not do so, you do not exist for them.

Yet existence defines not only individual living beings and individual details, concepts, events, and circumstances, but existence defines entire environments and realities just as well, along with entire oneness altogether. Existence defines everything interacting directly as an entire environment. Furthermore, existence defines everything interacting objectively directly or implicitly as an entire reality. While existence defines everything interacting objectively, subjectively, highjectively, directly, and implicitly as a oneness, and these can include entire worlds and realities by the zillions, as it is the case with the One, or Life, or the wider world, or the Supreme Being, you may refer to it in any manner you wish, understand, and are allowed.

Existence defines everything, separately or as a whole. Your physical body exists objectively, your life exists possibly eternally, your intelligence exists subjectively, your one exists, your interconnectivity exists, your world exists objectively, our higher world exists highjectively, and your soul exists highjectively. Your oneness or lifeline of existence exists at all levels, since your existence defines your body, life, and intelligence simultaneously in a similar manner, and therefore it defines your oneness. You are who you are. More precisely, you exist as a oneness, mind body and soul. And when you state that you exist, you state simultaneously that you are your physical body, that you are alive, and that you are intelligent. You are who you are. And this is exactly why you use the word 'being' to define yourself and anything else in this world, because you seem to define yourself the most through the existence of your oneness or lifeline of existence, of your life, of your intelligence, and of your physical body.

But let us take some examples here, in order to understand yourself as a lifeline of existence. You play your favorite videogame "GTA5," as Michael. All videogames are entire computer realities, many times correspondent to this world, for an increased experience and credibility. While "GTA5" is an open world videogame, offering you the entire replica of Los Angeles to drive cars, walk around, run, jump, and interact with thousands of people as you wish, as you do whatever you want there. It is a genuine inner computer reality, correspondent in many details with the real Los Angeles. The current computer interface is rather limited, as you know it well. But even so, after playing for hours, you actually become your videogame character, making it part of your lifeline of existence, at its end.

And as you study yourself closely now, you are the inner self from your inner replica of the world, embodying the physical body, and then embodying again Michael from Los Angeles, which is your last body or self from your lifeline of existence. But preceding all these selves is your soul or higher self. Which means that right now, your soul plays "GTA5" through all of you, as all of you.

And this is your lifeline of existence, since you are always mind body and soul as one. And more, since you are your selves of your lifeline of existence besides your soul, inner self, and physical body, since your soul may have souls, in larger or smaller numbers, depending who you are. And if your souls can reach Life herself in the beginning of your lifeline of existence, then you are divine, with your entire lifeline of existence divine.

But let us study more your lifeline of existence. You have more inner selves on your lifeline of existence, as the first reflexive inner self from your basal ganglia, your second intuitive inner self from your second, middle reptilian brain, and your third intelligent inner self from your third brain, the human cortex. Yet among all these inner selves, you identify yourself the most with the third, intelligent inner self from the left prefrontal lobe of your cortex, since this is where you live. And you do so since only your intelligent inner self has an intelligent awareness. And now this is your inner self.

Let us see now how you reason. Because you walk around with your legs, you see with your eyes, and you speak with your mouth, which means that you reason with your brain. And the current science looks very hard in your brain to see how you reason, because since you use your legs muscles to walk around, it means that you must flex some lobes of the brain in order to think, and now this is what science studies. Which makes sense, but you never manage to understand how you reason.

Besides, you cannot exactly flex the lobes of the human brain to think and feel, so how exactly do you reason? Currently, science tries hard to decode the axon potential, as you decrypt files throughout computers or as you decode radio waves to intercept phone calls, and it does not work. Because the current science is not too capable in understanding anything.

Because you do not think and reason as the physical body or as the brain that is part of the physical body, but you think and reason as the inner self. Your intelligent inner self jumps, walks, and handles objectively all concepts and conceptions in the inner replica of the world throughout intelligent reasoning and throughout the entire intelligent cognition, and this is what you have to consider.

Because within your inner replica of the world, everything might be subjective in nature compared to this world. However, everything is objective and material in every reality where you are and for as long as you are there. Which means that, from the perspective of your inner self, everything is objective and even material for your inner self and for the multitude of intelligences inhabiting the inner replica of the world alongside your inner self.

As a reference, all replicas of all the people that you know in this world, in the family and in society in general, are now genuine living intelligences correspondent to the real people from this world. And now, all these act and react within your inner replica of the world as they please, but more importantly, they do so in any manner relevant to your cognition.

Since this is how you think and reason intelligently, through comprehensive intelligent mental models correspondent to your subject of reasoning in this world. You enact or reenact specific circumstances of this world, in your inner replica of the world, throughout entire mental models correspondent to this world. Yet you do so mostly ahead of time, in order to predict the future, or you do so in parallel with the real circumstance, and in this manner you reason, act, and react in this world, while superimposing your entire inner replica of the world on this world itself. And this is how you live your life, mind, body and soul as one. You walk with your legs, you see with your eyes, and you speak with your mouth as the physical body, while you reason and feel with all concepts and conceptions of your inner replica of the world as your intelligent inner self, while you experience everything as your soul, intervening or not here on Earth, as you please.

And this is how you have your understanding of everything from this world, every object, subject, event, or circumstance of this world. Yet you do so mostly in the real society, through social mental models in your inner replica of the world, mostly having you at their center as the inner self, since most of your social circumstances have you at their center, concerning only you.

Or concerning your loved ones, depending on circumstances, or concerning the entire world just as well, or specific concepts, as existence itself. Since while reading this book, you form an entire intelligent mental model of existence itself, as you reason independently alongside this book. Because I never state explicitly how to mental model existence, but I only trace its main details in the most convenient manner for you to be able to mental model existence in your own inner replica of the world, in order to be able to learn existence as comprehensively as possible. And if I had more patience to detail everything at a higher cognitive resolution, then yes, you could mental model more.

And this is how you reason, as your inner self. Because you cannot reason as the physical body the way your physical body walks around, because your physical body is in a different reality, in this world, and cannot access the intelligent conscious mind from its world to reason.

Which means that you can never decode the axon potential in order to see a memory of grandma playing the piano there in your mind, because neurons are significantly more complex than simple encoded information. Since neurons and entire neuronal networks hold entire inner worlds and realities, and there is where the memories are. They live there, as entire living concepts and conceptions in entire correspondent inner worlds, and you cannot decode these with first level algorithms.

Because life starts at the second animal intuitive level, with many of your conceptions intelligent, at the third intelligent level, and you cannot decode these. Similarly, you cannot form the necessary computer code for second level intuition and for third level intelligence, with your first level algorithmic computer code, for lack of correspondence. It is even ignorant to attempt to do so.

And this is the case with everything and everyone in this world, since everything exists, is alive, and is intelligent, every living body and every object alike, as cold and inert as they may seem, everything is full of life, full of any form of life, just because the field forming and defining all lifelines of existence everywhere spans everything, every being and object alike, in all forms of life, generating in this manner the various forms of life in this world: ionic, molecular, cellular, organic, plasmatic, crystalline, and nuclear. Life spans all these, since these are her forms of life. While what distinguishes these forms of life one from another is not only their appearance, but also their pace, because all forms of life allow intelligences to experience life at various speeds and resolutions, from the very slow crystalline form of life, to the very fast plasmatic forms of life found throughout all stars, nuclei, quasars, and galactic centers. Yet despite all difference between them, you will always be able to project into all mediums of the field, in order to experience there echoes of your own lifeline throughout them temporarily, and you will always experience these as quickly and as slowly as those specific mediums allow you while you are there.

But is there a relationship between Life, Intelligence, Interconnectivity, wider world, the One, the Supreme Being, and existence? Yes, certainly. Time, space, and matter or radiation define you to exist objectively here in this world, because these three are part of the continuum of our world. To go further, you can never define time, space, and matter or radiation in any way but through themselves, because they are at the base of everything in this world, since they are the continuum itself of this world. And since only existence is below them to define them, you may define time, space, and matter or radiation only through existence. Time, space, and matter or radiation exist and form the continuum allowing and defining everything in this world to exist. In contrast, the other three concepts: life, intelligence, and the physical body define you to exist in every existential form in the wider world, here and beyond our world. Life, intelligence, and objective bodies are related to time, space, and matter or radiation here in this world, and I will continue this idea throughout this book series, since it is comprehensive.

What can there exist in this world? Everything, obviously. The question is what there can be that does not exist in this world. The rest does not exist, and this includes fiction, stories, consensus, beliefs, testimonies, and laws, along with anything that you may ever dream and imagine. We are going to keep our research of the inexistent flexible from the beginning, while we may still define everything, even the fictional and the consensual, since these exist through people and through their own subjective, cognitive, and digital creations. While the consensual exists through people's agreements, and it even forms the first level of existence, the ordered or the consensual, throughout an entire Consensual Matrix.

There are eleven levels of existence. The zero level is the nonexistent itself. Note that you do not exist for everybody else even if you happen to exist on your own, for yourself, with no interaction with everything and everyone else. You are at the zero level of existence in this manner for the rest of this world, and therefore you are the nonexistent. Similarly, if you spend your life doing anything else instead of fulfilling your needs and meanings, as taking drugs or watching TV all life long, you are nonexistent for Life, for this world, and for everybody else, and therefore you are at the zero level of existence. And in general, I refer to this as the zero addicted level, since addictions are very common. This is why even your family leaves when you are at the zero addicted level, because you do not exist for them anymore while addicted, as you focus your entire cognition, feelings, thinking, life, and existence on drugs.

At the first, consensual level of existence, you exist through consensual orders and duties, on behalf of others, but not on your own behalf. Therefore, you still do not exist for Life on your own behalf at the first consensual level, yet you still exist for Life and for this world implicitly, through the masters, orders, lords, ideologies, or jurisdictions that you serve.

The second level of existence is intuitive, characterizing lower level living beings and intelligences alike, as all plants and animals. Or as all underdeveloped people, since in general, if you live your life focused on physiological matters, then you live your life intuitively at the second existential level. You exist as an animal for Life, society, and the rest of this world, and this is the second animal intuitive level.

The third level of existence is conscious and rational, characteristic to humans, if they ever reach this level and do not remain at the zero, addicted, irrelevant level of existence, or at the first ordered, ideological, hierarchic, consensual level, or at the second, animal level.

And so on, there are seven more levels of existence above, characterizing the higher beings, higher realities, higher causality, higher circumstances, and higher achievements of Life and the wider world. With the tenth supreme existential level defining Life, Intelligence, Interconnectivity, wider world.

Yet there is more to consider, since there is an existential correspondent continuity between realities throughout the wider world. Therefore, all realities of the wider world are not isolated throughout the wider world, but they are linked through you and through all living beings through many details and laws, as higher laws, natural laws, supreme laws, and natural details. In this manner, everything that you know may be distinguished as a consensual belief or an accurate fact, depending on circumstances, just because everything from this world and from all realities inner or outer relates directly to the higher laws, natural laws, supreme laws, and natural details standing at the base of this world, and even to the higher worlds forming, holding, and maintaining this world.

In this manner, accurate facts are based on more specific accurate facts, which are based on more specific accurate facts, up to supreme facts found at the base of mathematics and classical physics today, at least, which are part of the natural laws or natural details of this world. While all statements unproven by accurate facts remain as simple beliefs, yet they may still be accurate facts themselves, depending on beliefs, only that we fail to understand them in a rigorous manner, for lack of permission or knowledge. Accurate facts may be easily understood and proven right or wrong through any accurate knowledge including theorems, while beliefs lead to debates and to entire ideologies of beliefs. These may end up in enslaving entire groups of people, entire nations, and entire worlds. And if existence manifests beyond this world and beyond this life, then you end up with people and intelligences enslaved indefinitely, just through the lack of rigorously understanding existence, life, and the entire world in all its realities.

Therefore, you have two kinds of studies related to existence: a consensual one based on beliefs, and an accurate, reliable one based on accurate knowledge and accurate facts. Philosophy studies existence empirically, through beliefs or empirical concepts, and this leads to significant debates and to fiction. Do not expect to learn anything from these besides dogma and divertissement. Spirituality studies existence in general, along with other worlds and other lives, along with proper pathways that you should follow throughout your wider existence. This is certainly useful to know and apply in life, and the entire world could become a better place if anyone would only follow this pathway. Yet spirituality is also based on beliefs, on spiritual beliefs, while these may be accurate or not, and you never know without undergoing a comprehensive, rigorous, accurate study of existence and of this entire world. And it is the same with religions, cults, hierarchies, jurisdictions, and consensual, hierarchic brotherhoods, since these are formed of consensual laws and beliefs that they define and instate in a conflict of interests. All these consensual concepts and knowledge may still exist, but only subjectively, at the first, consensual level of existence.

Yet many times, your existence and interconnectivity within hierarchies, ideologies, and jurisdictions do not relate to accuracy or even to righteousness and goodness, but to material agendas, social status, and privileges, as these are at the first level lifestyle and development. Because once consensual knowledge, beliefs, and orders reach your cognitive system and manage to install themselves at its core, you judge everything through them, you really believe everything to be accurate while it is only consensual, as this is called indoctrination and it is very common. Or not, because you can always fake it within ideologies and jurisdictions, only to maintain your social status and material privileges.

What is the difference between the ordered, consensual existence of the first level, and the living existence of the second level and up? It always relates to the interconnectivity between living beings within families and society. Because ordered or assigned social interactions bounded by laws, rules, regulations, beliefs, and entire hierarchies, ideologies, and jurisdictions results in an entire world filled with identical living beings, since they think, behave, and interact similarly, because everything is ordered, but you do not have the necessary variety in thinking, ideas, judgment, and interaction to generate all the knowledge that you must have as an entire civilization to cope with the entire environment, and you certainly fail. This is exactly why robots and computers can never manage to live an entire life on their own in society, since they lack cognitive diversity and genuine, unique, successful ideas. In contrast, the living existence found in all living beings and intelligences is based at least on intuition, as it is the case here with plants and animals, as intuition and intelligence are capable to offer unique solutions to all problems while these living beings and intelligences fulfill their needs. Yet intuition is only the second level thinking, followed by the third level analytical, rational, intelligent reasoning specific to humans, followed by the fourth level higher reasoning specific to angels and superheroes, all the way up to Life herself, the One, the wider world, Intelligence, and Interconnectivity, at the tenth supreme existential level.

What we want in our model of existence are accurate facts. Do other worlds exist? Where? And how do you get there to see them for yourself? Now prove it. Because simple beliefs are insufficient in helping you understand the wider world in all its details. As stated, it is the job of science to study, understand, and explain existence accurately, yet science persists to claim that what exists exists and what does not exist does not, and nothing else. Well, science does not even state this directly throughout its scientific studies, but only implicitly, because science never studies relevant topics as existence.

It is easy to remain captive within worlds as this one through ignorance alone and nothing else. Imagine a place used to imprison the most capable beings in this world, without walls, but only through ignorance of the outside world. And now, with science defining everything consensually and not necessarily accurately, and with science also controlled minutely by those controlling this world, as a social tool in their hands, you live your life in ignorance, without ever becoming aware of what exists out there.

For example, somewhere throughout the prisons of South America, if you happen to become pregnant, they keep you in a specific area of the prison, along with all mothers and children, in prison, many times for years to come. Study now these children born and raised in captivity, to find them remarkably different than the rest of the convicts in this world, since they are happy, they play everywhere, they run and laugh full of life as they play and interact with the rest of the children there, as they seem to maintain their freedom regardless of their assigned captivity. While their mothers are unhappy with their prolonged confinement.

Why are the children happy? Because their existence is

limited to those walls and tight cells, with the rest of this world being only some stories that they hear, but not their actual existence. Everything relates with the human condition, available since birth, regardless of its characteristics, and now this is what you have.

Because what you see is what you get and now this is the human life, keeping you happy. Only that, if you remember more than what you see, since you had it before, and now you do not have it anymore, it keeps you unhappy, since you always desire to have it back. Because your own intelligences send you specific developmental and environmental needs to have it again, to achieve it again in your life, freedom itself, along with the entire free human environment, and you cannot achieve it in prison. And so they punish you for your failure to fulfill your needs and meanings, keeping you unhappy and in pain. While if you are born in prison, then your intelligences are relatively content, leaving you alone, and not bothering you too much for your confinement, since this is your actual life.

To study this example further, all convicts crave for freedom while incarcerated, and as seen, they do so only because they are aware of a different existence beyond the walls of their prison. Once they lack this knowledge or awareness of a different outside existence, through amnesia or birth in captivity, the outside existence becomes inexistent now, or even fictional, with everything surrounding them as their direct, objective existence, unconditionally. Since this is the difference between an environment and the entire reality, both defined by the same existence, only that existence defines the environment as everything interacting directly, while everything interacting with you directly and implicitly is the entire reality. And you may go further to consider everything interacting in an additional subjective and highjective manner to be your wider world, or the One. And this limits your wider existence, since these are the only types of interactions that you can ever have.

Why exactly are the mothers unhappy? Because from a cognitive perspective, there are two incarcerations taking place simultaneously. One is extrinsic, through the conviction and punishment that the judge and the entire society gave them, while the other is intrinsic, through your own beliefs and mentality about life and existence in general. Your extrinsic incarceration relates to your extrinsic need for freedom, a need to be free and go in the outside world at once. Just because this is how you understand the concept of being free, to be outside and to fulfill all your needs outside, along everybody else. Because throughout society, the right for freedom and implicitly the need for freedom are defined and propagated in an idealistic manner. In contrast, your intrinsic need for freedom or your intrinsic need for being in the outside world relates to your intrinsic need to fulfill your needs, regardless of where you find yourself. In this manner, intrinsically, the prison environment suffices to fulfill your needs, as long as you live your life at the zero addicted and first consensual developmental levels. You have food, drugs, water, shelter, proper natural environment, and sufficient company from convicts, perfect to make a living, even an abundant living if you know how to limit yourself.

But now, with children born in captivity unaware of the idealistic concept of freedom, just by being able to fulfill their common needs, and by being around their friends and loved ones continuously, surrounded by relative plenitude, safety, love, and many times by harmony, this becomes all that exists for them, as it suffices to make a living and to grow up there, even harmoniously. Because they are never ordered and constrained at the first consensual level, but their mothers are. While the children are actually free, even behind walls. Just as you are free, even behind the tight confinements of this world.

There are rumors about this world being a prison planet or a purgatory. And this might be the case, as long as you suffer of amnesia regarding any wider life and existence that you might have out there throughout other realities, eternally or not. You love your food and drugs here on Earth, along with your company, shelter, safety, job, and brotherhood, so who cares about inexistence and what it might be and mean, as long as you always have everything that you need? It might be the same with coma, sheer servitude, and strong addictions, while all the loved ones outside your tiny world struggle to wake you up to this world and get you back to who you are, to your wider world, and to your wider existence, before it is too late and you expire.

What exactly may ever keep you captive within tiny, inner worlds? How can this be possible? Through ignorance alone? Is there a high wall at the end of this world keeping everyone ignorant of what it is beyond? Are there strong walls defining what exists from what it does not exist? No. Everything that exists is always around you, while everything that does not exist does not exist and therefore is nowhere, not behind tall walls, nor behind anything, but it simply does not exist. However, by altering the definition of existence itself, you may narrow the amount of what it exists, making everyone assume that what still exists in this world beyond people's awareness does not exists. There is an entire developmental level above what everybody has in the current human society, the third, intelligent human developmental level. But nobody cares, since it is never mentioned, defined and implemented by the media, science, education, and entertainment, and this is how many people live life underdeveloped throughout addictions, servitude, and low animal instincts. Since as stated above, it does not take walls to make the distinction between what it is and what it is not, since the simple definition of existence suffices to do so, followed by the simple ignorance and enforcement to believe it.

For example, how can you ever tell that all lights in the night sky exist objectively, and that they are objectively there, including the Sun, the stars, and the Moon, if you can never go there to see them for yourself, or if you cannot have any accurate scientific proof of their existence? Because whatever science and NASA offer is only a mockery, yet no one cares.

Because science promotes beliefs, not accurate facts. Because science does not prove its statements through accurate facts, but it does so through consensual reference to research done by other scientists, who base their scientific studies through further references, all forming the basic scientific consensus that you know well.

Are you pertinent in understanding the concept of existence, and through it, are you capable to understand and explain everything in the wider world and in all its realities? Well, tough luck now joining the scientific consensus, since none of these scientists will ever refer your work on the back of their books and research papers. Science will claim indefinitely that what exists exists and what does not exist does not, or that everything exists probabilistically according to quantum mechanics. Or that this world is the only reality, and this life is the only one you have. And so you stay here in this world indefinitely, or you go from here straight to the promised higher realms of your own ideologies and jurisdictions.

But now, what if you can prove that your dead grandma still exists, since she still talks with you now, days and months after she died? You even have an accurate proof of all these, since you hear her well every day. This happens to people sometimes, whenever their loved ones pass away. Spirituality claims that these are still around, sometimes unaware that they are dead, but how do they exist without their physical bodies? And since you are the only one hearing them, does it mean that existence itself is more complex than what science claims, if science ever claims anything regarding existence?

There are two circumstances, since either you answer your grandma, or you do not. At the same time, either you tell your doctor about it, or you do not. If you do, then you will have to take medication from then on, because there is this sickness that may develop in your brain, as tumors, aneurisms, and many others, which always start with your grandma. You take your pills, and you never hear dead people again. You are cured. While about that tumor, it will appear in your brain just as the doctor had predicted, it grows faster and faster, and unfortunately, it causes your death, after it took away your higher ability. From the medication itself.

Because your own knowledge regarding existence is very limited. Not that you will ever know what it was all about, and not that your doctor knows anything about the medicine that he prescribes to you, what it contains, and what it does to your brain and cognitive system. It says so in the medical system, which he must give you this medicine, and this is what he gives you. He does so every time people start hearing voices or have visions, he gives them the medicine as stated, and it works, since these stop. But then, all these medical complications occur, and unfortunately, the people die. All doctors do the same, they have their bonuses and promotions, and this world becomes a better place. For the doctors.

And if you ever wonder what they put in your food and drinks this entire time, it is similar medication, they do so just in case you develop higher abilities, they use an entire cocktail of drugs and medicine meant to keep you isolated entirely in this world, while keeping you weak, heavily sedated, sick, and in pain, while keeping you always on your way for your terminal illness. And many times, this is why you have to drink, smoke, and take drugs, to be able to escape the continuous pain and discomfort that these additives cause. And it is never a choice of discarding your medication or switching to organic food and now grandma is back, because your medication and additives are meant to amputate your higher abilities indefinitely, irremediably.

And when this is the case, then yes, what it exists exists and what does not exist does not exist, as science has stated this entire time. And it is the same with the multitude of vaccines that newborns take, just in case, because there are souls out there that come here and may be too developed for you, allowing you in every manner to see behind the veil, while those tyrants found throughout the upper social layers do not want this.

And if you do not tell anyone including your doctor, then you may answer or not to your grandma, and you take it from there.

But is grandma actually here, or it is only your imagination?

You will never know, as long as you cannot understand life, existence, and this world. How exactly are you now supposed to know what to do when you also die? There are promised upper realms to go to, there are your dead loved ones waiting for you, none of these may take place now since you are in the wrong reality, you are alone and highly vulnerable now when you die and ever after, so what can you do?

Your religious and spiritual beliefs will certainly help, and these bring you right back for another life here in this world, or they send you to specific promised lands anywhere else, larger or tighter. Or they help you exit your wider existence altogether. Which one is best? Are these promised lands genuine or only lies? It is a sin to think so, and many times, it is a sin to reason at all. So you have to find a way yourself, but you might want to learn and develop more before you make your choice, since these kind of choices tend to be permanent, eternal, forever, or this is also what they claim.

But how exactly can you decide what to do, right now, if you are too ignorant about the wider world and all its beings and realities, and if you are too weak and too disabled from drugs, medicine, and additives that you have been taking this entire life, ruining your cognitive system? Mostly if you are indoctrinated with beliefs specifically implemented within your cognitive system this entire life, in order to render you now to behave in very specific manners, for the rest of your eternal life. It takes an effort to reason independently, as it is always your own effort.

Can existence be distributed equally within distinct worlds, with nothing in between? Throughout my studies, I refer to all worlds, realms, and distinct places of any kind as realities. Existence has various natures, levels, scales, and lines, among many other attributes and characteristics. If you fail to understand these, it will affect your entire reasoning and awareness throughout life. This does not mean that once you understand existence for what it truly is, you start seeing dead people all around. Or this may also be the case, depending on your higher abilities, yet you will start understanding this world in its wider complexity, and this is development.

Will your cult, hierarchic brotherhood, school of thought, and religion banish you if you start suddenly to use your reasoning? Yes, certainly. Therefore, it is safer to remain within their beliefs throughout life, to do exactly as they say, and never use your own judgment and awareness. Everybody does so, and so should you. Who would ever want to start seeing dead people and monsters all around? Who would ever want to start understanding dead people and monsters throughout life anyway? Not you. Yet when these specific monsters happen to be your creation, you might be interested to see and learn what you bring in this world, through your own beliefs and behavior.

But what else can there be out there beyond the common existence, important enough to study and understand? There is nothing anymore for you, if you ever took the medicine, dogma, vaccines, and food additives. While if you never took any of these, which is impossible, then you could certainly be capable to take pictures and videos of ghosts then, as you see happening throughout the media.

Yet what do pictures and videos have to do with all these, if they relate with the existence of this world, while ghosts relate with the existence of other realities? The two are incompatible, those cameras will never capture anything, yet people never care. This is the way science works, by spreading its beliefs in every manner, even throughout pertinent scientific documentaries of ghost hunting from famous scientific TV channels, where people take pictures of ghosts season after season with no significant results. While millions watch the shows, drink their beer, and remain diverted.

If taking pictures does not work, how about hearing dead people in your head as it happens with your grandma? In fact, if you hear dead people with your ears, this is also impossible, since your ears are meant to sense the real, objective vibration in the air around here in this world, as all these are part of this world. When you die, in general, you exit this world, since you cannot interact with your avatar anymore, which was your physical body. Or this is usually the case, yet there are highly powerful intelligences out there that can do anything. Or almost anything. However, if what you hear manifests directly in your cognitive system, in your head as it is said, this might be something, since your cognitive system has three parts: subjective, conscious, and highjective, all capable to interact with other realities, inner and outer. There is more to clairaudience and clairvoyance in general, and I study it throughout other books of this series.

Yet you do not have to wait until you die in order to reach the boundaries of this world. You may project throughout astral planes, you may lucid dream, or you may dream normally as you do every night. Since all these take you out of this world, into other realities. And every time you do so, the existence of other realities becomes primary and objective in every manner, with everything else not to exist elsewhere, for as long as you are there. This is why you never realize that you dream, that you are not in this world anymore, regardless of how different everything may happen around you there. And this is the case just because all existences come with their own worlds, beings, lifelines, and mindsets. In fact, you were not even supposed to be in another reality, since all realities are tightly confined, with nothing else to exist outside them. But you happen to reach other realities now only through yourself or through your own lifeline of existence, through your own projection, just because you are a multidimensional living being yourself. And this is the case with all living beings, human or not.

In other words, you switch realities occasionally or continuously, just because your own existence is more complex than what science may state today. And now, with you lacking awareness of your complex existence, it is as owning this large mansion, with you living in the bathroom the entire life, unaware of the rest. Who does this to you? Since it happens with everybody, Masses and Brotherhood.

The concept of existence is by far more complex than what can fit here, which is by far more than the trivial statements 'to be or not to be,' and 'what it is is and what it is not is not.' Let us study now objective worlds and realities.

2 OBJECTIVE REALITIES

By definition, realities are, include, represent, define, and span everything that exists objectively. There exists nothing outside our own world, otherwise it was part of our world, making it larger. How can we ever consider other realities under these circumstances, if other realities do not even exist? Everything happens through you, since many times, you are more complex and more meaningful than entire worlds and realities. Because you are an entire lifeline of existence, and you have within yourself a zillion inner mind realities, while you are part of other higher realities just as well. In this manner, everything that exists here in this world is objective in nature. Everything that exists in your inner, mind realities is subjective in nature from our perspective, while everything that exists within your higher realities is highjective in nature for you, because existence has three relative, correspondent natures: subjective, objective, and highjective. And when you summate all realities of all living beings and intelligences, you obtain the One, or the wider world, or Life.

As you already notice, Life does not exactly inhabit the wider world in all its realities through all her living beings and intelligences, but Life forms all realities of the wider world through all her beings and intelligences, and this is how she lives her existence. And this is how Life is the wider world in all its realities, which she makes and maintains herself, to the point where you cannot distinguish Life from the wider world and from Intelligence, Interconnectivity, the One, and the Universal Mind, all being One, all being interconnected, while they exist. And I refer to the existence defining them as Existence.

While Existence defines you too as being alive, but only when you fulfill Life and the wider world. Because when you remain diverted, unfulfilled, and underdeveloped, you do not contribute to Life, you do not count for Life, you do not exist for Life, and therefore Existence cannot define you that you exist. You are inexistent or dead, regardless if you are still alive for yourself. And this is exactly how Life removes all living beings who fail fulfilling her, who are unwilling, too old, or incompetent.

Yet there is more to study here, since existence can be real and consensual. Through your consensual existence and therefore consensual fulfillment as this takes place throughout the consensual hierarchies of the current society, you do not fulfill Life and the wider world, but you fulfill specific people controlling this world through you, or controlling you altogether. This entire social consensual system is organized minutely, with all the necessary documents, stamps, demands, and signatures minutely kept. This organized consensual legal system spans the current human society, engulfing you, regardless if you are from the Masses or the Brotherhood, going further above and beyond to span and engulf most of the wider world. This is not Life but the Consensual Matrix, it is of the first, consensual, ordered existential level, and it uses you and everybody else on its own behalf, taking you away from Life and from the entire wider world, while killing you consensually and maintaining you consensually dead indefinitely. You are even considered legally a dead body, corpse, or consensual corporation, since only in this manner, you may remain within the Consensual Matrix, for existential compatibility reasons. Yet when you persist to state that you are the living human being, they remove you from the Consensual Matrix, since the Consensual Matrix never interferes with Life and the wider world, again, for existential and legal reasons.

As stated, existence defines the existent from the nonexistent, while Existence is the specific existence defining the One, or Life, or the wider world.

While as already seen, the existent may be direct, implicit, subjective, objective, highjective, fictional and consensual. All realities below this world are subjective, as all mind realities and computer realities. This world remains objective along with all objects and subjects that it has, while all higher worlds and realities are highjective in nature, all the way up to Life, Intelligence, and the wider world.

The difference between the fictional and the consensual is that the consensual is an entire abstract agreement among a group of people spanning an ideology, political party, hierarchy, jurisdiction, or entire consensual society, while the fictional is abstract and imaginary just as well, yet you do not have to enter in agreement with anyone in order to create your fiction, since you may do so on your own. While the nonexistent is nothing at all, and sometimes it is called the void.

Yet all realities of the wider world do not exactly float as gigantic spheres of existence into the nothingness of the void, just because there is no space or time in the void, since if it was, it was part of our world or of any other reality, and it was not the nonexistent anymore. Because all realities are formed through a modulated matrix holding them within an abstract form, and not in a specific continuum. The hardware of your computer does not hold your videogame physically or even electrically, but it holds your videogame reality in a digital matrix that takes no time and no space, because it is abstract and intelligent modulated, and that is the matrix. More precisely, you may see a modulated matrix from the outside, while from within realities, you see only a continuum and a field of possibilities according to the natural laws and natural details of that specific reality. Yet realities do not have an outside, since if there was anything outside realities, it was already part of them.

Even more, these inner realities do not even use energy to be sustained by their higher worlds. Your computer does not use energy to hold your videogame realities even when you hear the fan intensifying throughout more complex parts of the videogame, because all the energy that your computer uses does not go into the videogame worlds since it is impossible, because you cannot transfer objects, subjects, or information from one reality to another. While all the energy that you put into your videogames is dissipated as heat by your computer fan, and it remains here in this world. You are simply using your expensive computer as an electric heater when you play your videogames, with the same efficiency as an electric heater.

While as you notice, there is no space between the realities of the wider world, just as there is no physical space within the hardware of your computer when you run several videogames simultaneously on one computer.

And it is similar with your own mind realities, because you have a multitude of mind inner realities running simultaneously throughout your cognition, but there is no space between them, since they are not exactly spheres floating in the void, but they have their specific living cognitive matrices within the brain and the rest of the body holding them. While all intelligences of your cognitive system inhabit all these inner realities throughout your conscious and subconscious cognition as you perform all the inner and outer tasks and activities throughout the fulfillment of all your needs, inner, outer, and higher.

As stated above, there is something out there everywhere, and it is called the void. Yet words as 'something,' 'outside,' and 'everywhere' have no meaning outside our world, since nothing exists out there, and this includes our spacetime continuum. You cannot even have walls and physical boundaries defining, separating, or containing our world, since there is nothing separating this world from anything out there, since it does not exist. Besides, these walls have no meaning themselves, as closely as they get to the edge of our world. Because these edges, limits, and vicinities cannot be defined from within and from without.

And this is exactly the inexistent. But where exactly is it? Where and how does the inexistent start to manifest? Is the inexistent the Void? Do these questions even make sense, since the inexistent cannot exist itself, while these questions apply to something that exists? Is the inexistent even valid?

There is by far more to the inexistent, since the inexistent is relative to living beings and to their cognition and way of understanding, mostly. Because the inexistent comprises both the absolute inexistent, along with the inexistent related to ignorance.

Do all ghosts live in the Void? What is the Void? Because if the Void exists, then it is part of our world. But no, no one lives there, since it is impossible, besides some highly powerful higher beings, who manage to isolate themselves from everything, and live there on their own. Intelligences you will never have the chance to encounter, since they manage to isolate themselves from you. Yet you may visit the Void at night or throughout your astral projection, if you know what to do, and if you remain conscious while doing so. Because you may exit our world not physically or objectively by travelling from objective physical place to objective physical place, but you may exit our world subjectively and highjectively.

But where exactly do you astral project, if nothing exists beyond our world? Nothing exists objectively beyond our world, but everything exists subjectively and highjectively below and above. And you live there, in your other realities, through your other selves and intelligences, because this is where you go and what you become when you dream, daydream, and project.

There are realities out there, countless in number, none exists objectively, and this is a good way to start understanding everything. Daydreams never take place within our world, since that would be called direct manifestation, bending reality, or mind over matter. Because our world spans everything that exists objectively. And if your dreams could manifest anything objectively, then they become part of this world, with your dream worlds also part of this world.

While objectivity is significant here, because by definition, realities span and include everything that exists objectively. There are other realities that exist subjectively and highjectively in rapport to our world. And this is what science ignores, the fact that the subjective and the highjective are relative, correspondent natures of existence and therefore still part of existence, similar to the objective existence. Your videogame character exists, since he is right there on your computer display. You have been playing him for months now, and therefore he cannot be part of the nonexistent. That entire computer world also exists, yet it exists only subjectively, and not objectively. Since if they existed objectively, then you could go there yourself, within your computer worlds, and play the videogame for real, in person. Yet in that case, your computer world was part of this world itself, since this world spans everything that exists objectively.

Your computer character exists subjectively, and he is your avatar or computer self within your computer world. For your character in the game, if he was ever able to think, you also existed here in this world, but you existed highjectively for him, not subjectively, nor objectively. Let us see why.

All realities are objective in nature when you are within them, when your perspective is in the same reality. This is why you take all your dreams for real, because you are there, within your dream worlds. While as long as you are there, and as long as your perspective is there, everything around is objective and credible. That is your world then, and nothing else exists objectively and materially besides your world.

It is almost similar with computer worlds, since after months and years of playing the same game, and when you play it for sixteen hours straight, everything becomes objective in there, that is your world, and there is where you live. This is a matter of dissociation, and it happens because you relate more not to that computer reality, which remains subjective continuously, but you dissociate by remaining within the replica of that computer world from your own mind. Because it is in your mind where your character from the game becomes alive, you give him life there, through the specific inner intelligence that you use in order to interact with him through the display of your computer.

And now you have three realities to consider: this world, the subjective reality of your computer game, and the mind reality created by your brain in order to help you think and react throughout the videogame. Three realities, three separate existences, one objective, and two subjective.

You always sense and reason through mind realities, not only throughout videogames, but every time throughout your life. In a way, you may state that you do not interact exactly with this world, but since your understanding about this world is in your mind, in your own private inner reality, which is an inner replica of the world, and since you use this inner replica of the world in order to interact with this world, then you interact directly with this replica of our world, and not with our world itself.

And this is where problems arise, because you never reason and behave according to this world itself, but according to your replica of the world. Which cannot be accurate, since you cannot transfer accurate information from one reality to another, but only personalized copies of information. Throughout life, expect to be wrong many times, and expect to learn from your mistakes, since only in this manner, you may revise and update your inner replica of the world. Because, as stated, the inexistent is not only absolute, but it is always augmented by your own ignorance and lack of awareness and understanding of the existent around, transferring it in this manner to the inexistent, all these accounting for the lack of completeness and accuracy of your inner replica of the world.

Therefore, for you, the inexistent remains outside your own knowledge, awareness, and understanding, which remains outside your own inner replica of the world. And you can never acknowledge the inexistent, just because you have to use your own reasoning through your own replica of the world, in order to assess and understand the inexistent. And since the inexistent cannot be there within your own inner replica of the world, otherwise it was the existent, now you can never acknowledge it nor comprehend it. Therefore, you can never know what you currently do not know, simply because you ignore it. Later on you may find out, by using your updated, future replica of the world in order to assess and understand your past replicas of this world, and the difference of the two is the inexistent that you have managed to understand and fill in ever since.

Mind worlds of all types are also subjective, and this is the case with all mind worlds of every living being. Because mind realities are formed, held, and maintained by brains found in this world. Therefore, it is this world holding the specific matrices of mind and computer worlds. All realities formed and maintained in this manner by our world are lower or inner realities, since they are held by our world. All our inner realities exist in a subjective manner, while everything existing in our world directly, exists in this manner objectively. This includes all rocks and objects, all plants, animals, and physical bodies, all water and air, all solids, gasses, and liquids, all waves and vibrations, and all fields and radiation, including visible light. Throughout this chapter, we focus on this objective existence of all realities, since all realities have an objective existence when you are in there, part of them, and we will use this perspective in order to understand both the existent and the inexistent. This means that we will use the objective perspective in order to understand everything else within and beyond objective realities, just by moving our perspective within these realities.

As seen, existence has three natures: objective, subjective, and highjective. These are relative to each other. The objective existence characterizes you, along with the entire world and reality having you. The subjective existence relates to all inner realities of your objective reality having you, as your mind realities and your computer worlds, and these count in zillions. Even more, all inner realities have inner realities themselves, which have inner realities, as all these exist subjectively, since they are inner realities.

Similarly, this world may be held in a matrix by our outer reality. If this is the case, which seems to be according to many religions, brotherhoods, schools of thought, and individual testimonies, then our higher or outer reality exists highjectively for us, not subjectively. You can never understand our higher reality in an abstract manner as you understand any concept or circumstance from anywhere around or from any computer world, but you may understand it only highjectively, since this is how our higher reality exists. While you always need a higher reasoning to do so, along with additional higher knowledge to perform all your higher mental models throughout your understanding.

How exactly can you understand our higher reality? The human mind has three major intelligences: subconscious, conscious, and highconscious. You may understand everything highjective in nature through your highjective mind, while you may perceive and understand everything objective around you through your conscious mind. How exactly do you employ your highconscious mind? With care and talent, since your entire link with your highjective intelligence is severed in every manner, through underdevelopment, dogma, vaccines, drugs, and harmful additives. It is an entire enterprise taking place throughout this world and throughout life, meant to keep your highjective intelligence amputated from your cognitive system. This happens to everybody as it makes a considerable difference in this world and in your life, yet you do not notice it, since life has always been in this manner. And this happens to everybody, yet no one ever cares. And now, if they can amputate your conscious intelligence, which many times they do, this is a bonus.

This is how you may 'see' and understand your soul and its entire world and existence, through your highjective intelligence. As a reference, your computer character may also understand you in a highjective manner, since you and this entire world exist in a highjective manner for him. Yet he lacks a highconscious mind, since computer technology is in its infancy today. But if it was not in its infancy, if you used direct mind over computer interface, then his own highconscious intelligence was part of your own cognitive system, since they were linked together through the mind over computer interface, and he could understand you that way, directly, highconsciously. Or this is the case if you play the game in first person mode. Because if you play in third person mode, and if you want to keep your presence and influence secret throughout the videogame, then you kept your own mind separate from his, and he could not reach you directly through his highconscious mind. If you even allowed him to have a highconscious intelligence besides his subconscious and conscious intelligence, highconscious mind allowing him to have any awareness of his higher world, which is this world.

This does seem to be happening with you in our world. What a coincidence, but no, there are never coincidences in everything regarding realities, just because all realities are similar in structure, functioning, and purpose. Further up the line of existence, there are higher worlds and realities holding our higher realities, and this goes on upwards along our line of existence, up to the One himself, who is the entire wider world, everything that exists.

I define this direct relationship among realities as existential lines of realities within realities within realities. The entire wider world seems linear and highly organized in this manner, yet this is not exactly the case. The wider world is not exactly linear in realities, but it is shaped in trees and clusters more than linearly, since all living beings of any reality will always form inner realities throughout thinking by the zillions, inner realities forming similar inner realities throughout their thinking also by the zillions.

You notice how all these realities spanning the wider world are part of the inexistent, according to science, while they exist either subjectively or highjectively, depending on where you are positioned within the One.

Note how all these other realities exist not independently, but they exist through you, or they exist through all living beings in this world, as they exist through their cognition. Since this is the correspondent existence characteristic to Life. Therefore, there is more to you now than what science claims, since you are more complex than entire worlds and realities, since it is you defining, creating, and maintaining these throughout life, or this is the case at least for your own inner realities.

How do you do so? In many ways, but first, you must understand how realities in general are formed, held, and maintained. All realities have a matrix at their base, which holds their own continuum. The continuum of our world is the spacetime continuum, and it is held by a matrix, which is held by our higher reality. There are at least two different types of matrices used to form, hold, and maintain realities: natural as within brains, and digital and artificial as within computers.

The simplest type of matrix exists within and through the chips of all electronic devices, and it is artificially created by humans. Take your laptop as an example, while you are playing your videogame "The Sims 4." It is the motherboard of your laptop creating that entire virtual reality of the game, but not directly. The entire computer hardware exists objectively here within this world, and this includes the electricity throughout all chips and circuitry, since electricity exists objectively here in this world, because electricity is objectively real. The matrix itself is not the electricity, but the matrix is the encoded variation of this electricity. The matrix is the specific succession of ones and zeros, yet the matrix is not even the ones and the zeros themselves, but only their controlled, modulated, encoded, intelligent succession. This guided, intelligent, binary or digital variation of electricity, through its specific modulated succession of code, is the abstract matrix itself, which is capable and complex enough to form, hold and maintain an entire virtual reality, the inner reality of your computer videogame.

Do you see how there is no direct, objective transition from our world to the inner reality called "The Sims?" Since if there was a direct transition, then the two realities were one. Because it is the immaterial, subjective, encoded variation of electricity causing and becoming the matrix of the new inner reality, not the electricity itself. And this is the case even if there exists subjective electricity within the videogame, used by the sims. That subjective electricity in the game is used to power their lights, computers, and appliances of all kind throughout their game, but that electricity has nothing to do with the electricity from our world. Or it does, indirectly, as part of the field, since the field is everywhere in its intact form, and we have plenty of time to study the field throughout this book and throughout this entire book series.

The field is the matrix or the continuum of the One himself, it is the actual Universal Mind from spirituality, and this is why it becomes part of the continuums of all realities of the wider world. While you may experience the field here in form of the common electromagnetic fields, to help forming and holding all our inner realities.

And this is why all intelligences are compatible now, since they are made by the same variations of the field. And this is why you may project throughout all mediums of all realities, as far as you can get, because these are also formed and held by the same field. And this is why all forms of life are plasmatic at their base, because ions are held directly by the field.

Note that this encoded variation of electricity becoming the matrix holding an entire inner reality is not present within the inner reality itself, or it is not present there directly. This matrix forms, holds, and maintains not the inner reality directly, but it forms, holds, and maintains only the continuum of that reality. This continuum is what holds and maintains the entire inner reality. In a way, the continuum is the matrix holding the inner reality, as seen from the outer reality, while the matrix is the continuum of the inner reality, as seen from the inner reality.

In the case of our world, our continuum is the spacetime continuum, and this holds everything that exists here objectively, while also holding all matrices of all inner realities formed, held, and maintained here in our world, as these count in zillions.

And only a few pages ago, we could not even get out of this world with our awareness in order to understand everything else, but now we may see this entire wider world in its zillions of clusters of zillions of realities, with all living beings of the wider world holding these realities within and without themselves, while forming the One.

Your question now may regard the nature of the matrix holding our world. That may be of many types, as natural or technological, or as a combination of these two. Implicitly, our world may have a cognitive nature, as part of an extraordinary higher mind or brain, or it may be part of an extraordinary higher computer or higher simulation, or both. And this is the case because realities may be only naturally created or artificially created. While studying religion, spirituality, and countless of firsthand testimonies, it seems that our world is created, it is artificial, yet it still has a living cognitive nature, as being part of an extraordinary common higher mind. This extraordinary common higher mind allows higher beings to link to and come down in this world, as normal human souls, our souls. While they may have similar souls, coming from even higher worlds. And since they are alive, it means that their common higher mind holding our world is alive, making us alive as living human beings, while making our world alive. Including our world, part of the Universal Mind, which is the cognitive perspective of the One, and this gives our world its living, cognitive nature.

Why is our world cognitive in nature and still artificial? This is what records, beliefs, and testimonies state the most. Your own computer becomes a cognitive device even though it is still artificial, every time you use it in order to fulfill your cognitive needs. This is the case when you use your computer throughout your learning, development, social interaction, or throughout simulating objective events and circumstances from the outside world, as when you use your laptop through social media, or when you play your videogames online with your friends for months and years. Today's computer technology is still in its infancy, but wait a few decades, if humanity has a few decades left for you to wait, to find Facebook changed into a comprehensive virtual world, the Metaverse, capable to provide virtual gatherings and encounters throughout genuine artificial inner realities. If Facebook is still around in a few decades. Because everything happening within comprehensive virtual worlds or inner realities happens in order for you to fulfill your social needs directly, along with the rest of your needs implicitly.

Our world goes virtual very soon, and it seems that it does so through technology. And according to the majority of records, beliefs, and testimonies, from a higher perspective, it seems that this is what our world is, and this is how it functions. Because our world itself may be one of these highly developed videogame and computer interactions as "Facebook" and "The Sims" combined in one, in an extraordinary, entirely interactive social media platform where higher beings gather to interact and exist subjectively, higher beings that we refer to as souls.

And it is only through today's level of civilization that it seems that this world is going virtual through technology. But how else? Through higher powers, obviously, or through telepathy, more specifically. You do not have to link minds through computers, WI-FI, and networks, but you may link minds directly through higher abilities in order to form inner virtual worlds, as far as telepaths may interconnect. This is the natural way, for this world to think, imagine, and create together, if those controlling society only allow telepathic powers to manifest in this world, since they tend to clip them off everyone's cognition. While this seems to be the case throughout outside worlds, not only here. People controlling people throughout higher and lower social media platforms, how typical.

How do people control people? In every manner, as it happens throughout society, directly and implicitly, hierarchically and centralized, with and without their awareness. In what it concerns our higher world and ourselves in the higher world, control happens through ignorance. You lack knowledge of any of these, while you must know everything in order to escape their control. Because they never kill you as you see in all movies, but worse, they hold you captive indefinitely, while they use you, oppress you, exploit you, and harvest you systematically and minutely in every manner, throughout every moment of your eternal life. And this includes making you harm others in every way, including your loved ones. Through your ignorance.

What can you do? Overcome this ignorance. Because as we keep on learning throughout all these mental models of this entire book series, ignorance and underdevelopment in general are the quintessence of all problems in this world.

We are modeling realities from objective, material perspectives throughout this chapter, and this includes how realities are formed, held, and maintained. We have already done so for computers, and now we have to do the same for brains and living beings in general.

Your mind is the inner reality formed, held, and maintained by your brain, and implicitly by you. How do brains form inner realities? How do brains think? Let us now study, understand, and explain these two questions, since they are directly related, and we may do so simultaneously.

First, you do not need a brain in order to think, despite what science may claim. Protozoa are capable to interact with their environment, and they do so highly successfully, which implies empirically that protozoa think. This also implies that the protozoa are conscious, since they are capable to interact with their own environment. And they do so efficiently, because the protozoa will still be here long after the next cataclysm causing the death of humanity. Therefore, empirically, the protozoa are more intelligent and more capable than the entire humanity. And this may be the case even objectively and cognitively, not only empirically. But let us see now how single cellular organisms are capable to think, be conscious, and interact with the environment, without a brain.

All cells have cellular membranes. All that cells have to do is to pass through the cellular membrane nutrients, toward the inside of the cell, and to pass waste toward the outside of the cell, and nothing else. Some cells are also capable to move around in many ways, and this allows them to chase these nutrients, to chase their proper habitat, or to chase their existential niche, in order for them to fulfill their needs. Chasing, needs, and choices denote cognition or thinking, and this happens through normal thinking taking place within these inner realities, where it is made possible by specific cellular intelligences found within these inner realities, intelligences similar to your characters of your videogame "The Sims." Intelligences always do the thinking and care for themselves and for the physical body, while the physical body only contains them within their inner realities. And together, intelligences and physical bodies form life.

It is the same with you and with all living beings. Yet there is more to state here, since there are not only the mind and the physical body on your lifeline of existence, but a multitude of other selves, higher and lower.

The comparison is accurate, since this is how intelligences behave within inner realities, yet what it is not actually accurate is that cells and any organism has to form entire inner realities in order to think, and therefore in order to cope with the environment and subsist in this manner. What happens is that those intelligences are in fact in charge of everything, and they live their lives in charge of their physical body just as you live throughout life in your car while interacting with the road throughout your life. And even this is not exactly true, since from a nanoscopic perspective, when you study the protozoa very closely, you are not able to distinguish between the objective body of the protozoa and the actual inner world and intelligence anymore, both being one. The three of them are one: life, intelligence, and the physical body, as we have seen in the previous chapter, yet that was the case from a higher perspective, not from a nanoscopic perspective.

How does everything function? How are inner realities formed? How do living beings think? The entire cell holds subjectively its inner intelligences, by the zillions, intelligences specialized in all the necessary tasks to be performed throughout the cell. Study a cell now, to find it formed of a zillion cellular components, from ions, amino acids, and proteins, up to the nucleus and the mitochondria. But since these are ionic and molecular in structure, or have ionic components, they interact with the overall electric, magnetic, and electromagnetic field. And through this specific changing field distribution, all these cellular components are capable to form a zillion cellular inner realities.

While within computers, the electricity fluctuation was digital or binary, now within cells, this field variation is analogue and therefore relatively infinite in resolution, capable to form and maintain a matrix more powerful, faster, and of a very high resolution. This powerful matrix may hold the necessary continuum holding extraordinary worlds, which are full of intelligences.

You may assume that this is the case at the cellular level, but this is not so, since this is the case at a protein level, while proteins vibrate and move intelligently throughout the cellular field, and through it, throughout the overall field. Now multiply that a zillion times to get the cellular intelligence, since there are zillions of molecules including proteins within cells. Then multiply it a trillion times to get the power and capability of your own cognitive system, because there are trillions of cells in your organism. Because it is never a matter of the protozoa being more intelligent than you or than humanity itself when it comes to surviving second level environments as Earth, but it is always a matter of you and the entire humankind being determined and forced to fight and harm one another in order to extinct yourselves, even before the natural environment of Earth switches to the third dreadful level, becoming a threat to humanity.

In order for you to understand the inner realities of your organism, along with your cognitive system, and along with your thinking taking place within your cognitive system, you must first understand thinking and inner realities formed at your lowest level, at the ionic form of life. And in order to understand yourself at the ionic level, you must understand life in its plasmatic form, since you are of a plasmatic form deeply beneath your organic form of life. You may find genuine plasmatic forms of life within continuous fires, within stars, within the ionosphere, within the air all around, and within the empty space.

This does not mean that the essence of life is of ionic nature. Or this is the case, with this ionic life building up throughout cells and organisms one form of life on top of another, to become the living beings that we see everywhere. However, just because ions are the smallest structures that we may identify, this does not meant that they are the most elementary structures containing the essence of life, since life may exist and manifest at subnuclear levels, building up from there into larger and larger living communities in order to become what we have today. It is the same at the other end, in the macrocosm.

To return to ions now, the empty space is not empty, but it contains elementary particles, about one every centimeter cube. Since there is electromagnetic radiation coming from all stars, these interact continuously, causing the elementary particles to vibrate. Vibration means accelerated movement, and when these elementary particles happen to be charged, you have a direct interaction with the field, which is the well-known electric, magnetic, and electromagnetic field. You may sense this vibration in empty space as a very low temperature of the empty space, yet slightly above zero Kelvin, at around 2.72 Kelvin, everywhere around. And as low as this temperature or ionic vibration may be, you may still consider it plasma.

This vibration can never be chaotic or random as long as you have an intelligent vibration in its vicinity, because all vibrating particles throughout space influence themselves directly, through induction, through the surrounding field. Therefore, you may have any intelligent vibration anywhere in the universe, and this will spread out eventually to entrain and influence everything, giving to everything an intelligent rhythm, an intelligent life. Let us see how.

If these variations are intelligent, then the intelligent code that they maintain is enough to form a matrix, which is capable to form and hold inner realities as seen before. Intelligent life means an intelligence, a system of intelligences, or entire societies and worlds of intelligences, all living and interacting not in the empty space itself among the ions, but existing subjectively within inner realities held by the specific matrix formed, held, and maintained by the ions of the empty space. While the ions of the empty space vibrate intelligently and not randomly, not because they become intelligent themselves, but because the intelligences within the inner realities that they hold interact with their continuum in a living, intelligent manner, which interacts with their matrix in a living, intelligent manner, which modifies the vibrations of the ions in a living, intelligent manner, which modifies the field around them in the same living intelligent manner, which in turn modifies the vibrations of the ions in the vicinity and eventually all around, which in turn changes the shape of the matrix in the vicinity and eventually all around, which in turn changes the continuum of the inner realities, which in turn changes within the inner realities various structures, things, events, possibilities, whatever these intelligences are doing there within their inner world, in a living, intelligent manner.

And this is Life, at least at this lower plasmatic or ionic form of life. And then the cellular form of life stands right on top of the molecular form of life, with the organic form of life on top of the cellular, molecular, and ionic forms of life.

And to be even more specific, the ions themselves are not alive, while the intelligences themselves are not alive either, or not by themselves. The entire array of ions of the empty space is alive when also considering the intelligences that they form and hold within. The physical, objective array of ions being the physical body, and the groups, societies, and worlds of intelligences that they contain within their inner realities to be their system of intelligences, or their cognitive system. And once you have a physical body containing intelligences within its inner realities, you have a lifeline of existence, and therefore you have life. And in the case of the plasmatic form of life of the empty space, once the inner intelligences are capable to reach out and control the movement of their own matrix to further control directly the inner worlds held by this matrix, then you have self-aware and self-maintained intelligent life, capable to cope, adapt, and interact with any environment, and therefore capable to develop as highly and as adequately as necessary. And the same goes on within stars, only at a very intense level, since this is the plasmatic form of life.

There are not too many free particles throughout the empty space, yet by being so vast, it may still account for impressive inner worlds holding impressive intelligences. And since the plasma of the empty space is in direct contact with the plasma of all stars, with the atmospheres and fires of all planets, with the organic life, and with life of all forms, the intelligences formed, held, and maintained within the inner realities of all living beings of all forms of life may be connected directly if necessary, and this allows you to dream and project everywhere you choose. This is the case with this world, but you may do the same within all inner and outside worlds, since these are connected through the same plasma, same matrices throughout the same field making everything possible. And all it takes is an intelligent vibration to allow the necessary interaction with the field, and the field reshapes itself adequately in order to allow this intelligence to manifest, live, interact, and think, as it needs and as it desires. Since the field is the Universal Mind, and it is alive, it is the source of life and intelligence.

Can we sense or record this intelligent vibration in the empty space at least? Are we capable to distinguish it from random, chaotic vibrations? Yes, certainly, since the basic facts of physics allow us to expect random oscillations of ions as they interact directly with the electromagnetic radiation coming from all stars, as distant as they may be. And since this electromagnetic radiation is of all frequencies, you may expect a random vibration at about all frequencies, taking place at a very low amplitude, minus the lines associated to the common particles of the empty space. And if this is what we detect today throughout the empty space, then yes, the entire space is random and chaotic, and there is no life present. Yet if this vibration is coordinated by life and intelligence in any manner, as our mental model of the plasmatic form of life states, then you should notice and record an obvious high order in this random vibration, just as you can notice obvious cities and space structures within and around highly civilized worlds everywhere in the universe.

And yes, there is a difference between the natural oscillation that any elementary particle may undergo everywhere throughout the empty space. Measure it, to see it gathered at a specific frequency, a frequency associated with infrared radiation, with not much left on the rest of the spectrum. And this is the case everywhere throughout the observable empty space. Science refers to this as background radiation, associating it to the famous big bang theory. Study the background radiation and the entire big bang theory, to find them struggling to explain observable details independently, never offering a conclusive, accurate cosmological explanation. This is an indication that the big bang theory with its background radiation are not meant to offer a cosmological explanation of the universe, but they are meant to offer false explanations and interpretations to anything in the universe, leading to its living and intelligent characteristics.

And this is how you end up learning throughout your education of a dead, chaotic world that live in, with your own life and intelligence being simple anomalies in this void, lasting eighty years or so, around here in this specific tight corner of this world. While society and those controlling society cannot keep you more captive than this, locked up in the tiny bathroom of your extraordinary mansion, life after life. And you just love it. And you keep those around locked up in their tiny bathrooms, since this is what you have learned to do throughout life. And now you love it even more.

There are ions within all cells, in a substantial amount. These ions also vibrate in a same intelligent manner, forming inner cellular intelligences within your cells as they form it within stars, ionospheres, and empty space. Only that your cells are filled up with ions and charged components, making the entire model significantly stronger, faster, more potent, and more capable. This is your plasmatic nature, yet it does not stop here, since the organic form of life is in fact a plasmatic or ionic form of life at its base.

What characterizes cognitive systems is not necessarily their strength and amplitude through an increase in the number of ions, but the possibility of having a conducted, intelligent interaction between individual inner realities within the system, according to the needs of all inner intelligences of the system. Your brain does not light up entirely on the display of the EEG machine while reasoning, but it does so systematically, through distinct interconnected parts of your brain as they interact systematically throughout reasoning.

Now, by having intelligently vibrating ions in the field, you may create significant inner realities. Yet by having vibrating dipoles in the field, you obtain outstanding inner realities, of a larger cognitive resolution. Common dipoles within your cell are salt molecules, among others, and they are more efficient because they have one positive and one negative charge linked rigidly as they vibrate in the field, interacting more intensely in this manner with the field.

Why not having more linked ions vibrating in the field, for an even higher cognitive resolution? You cannot link ions throughout molecules in any manner you desire, since the natural laws of physics and chemistry stop you. Even more, the types of ions that you may link throughout molecules must have specific characteristics, and must be widely available in the environment. The most available elements in Earth's environment are hydrogen, carbon, oxygen, nitrogen, silicon, sulfur, and phosphorus. You may link these to form molecules only in a few specific manners, and out of these, only a few combinations are the most efficient while interacting with the field. And so you end up with the well-known dozens of proteinogenic amino acids used by organic life, with less than two dozen used by you throughout your body. Link amino acids further in specific manners and you form proteins, enzymes, RNAs, and eventually you form all cellular components, all cells, and entire organisms, by linking only these. Since this is the molecular form of life on top of the ionic type of life. And then the cellular type of life on top of the molecular type of life. And then the organic type of life on top of the cellular type of life.

Who exactly links these ions to form all these cellular components? Intelligences do, since as seen, intelligences of inner realities are capable to interact with the matrix holding their own realities. This is easier and more trivial than it seems, since intelligences only interact normally throughout their inner worlds, they interact with their own continuum as you do every day just by walking around, while the matrix adjusts itself in order to handle, support, and allow whatever goes on within inner realities. You may observe this change in the matrix held by your brain as it lights up on the display of the EEG machine throughout changes within your reasoning. You may also see and hear all changes in the digital matrix of your laptop according to everything going on in your videogames or computer software. And this is the case just because existence itself is the only one capable to move and transcend from one reality to another, by morphing itself at the threshold between the matrix and the continuum that it holds subjectively. This is the existential transformation from the objective to the subjective, and it is significant in understanding existence and the entire wider world.

You may consider the organic form of life distinguishing itself from the plasmatic form of life at the cellular level and above, if you choose to do so, yet this is not a structural characteristic, but only a reference. Because cells do not have to have a cellular membrane in order to allow them to function, as you may consider that the skin of organisms keeps their cells together in a larger bag, allowing the entire organism to exist and function in the macroscopic world, which is not the case. Cells are capable to live, think, and function without a cellular membrane, since cells are not only alive on their own, but cells are in fact communities of individual living beings. Cellular membranes appeared over one trillion years ago, when a significant change in the composition of the water of the ocean forced all living communities of molecules of that time to build and maintain a wall around themselves, in order to allow them to maintain the old water environment among themselves, keeping the new, toxic environment out. And this was the transition from the molecular form of life to the cellular form of life, with the building of the cellular membrane.

The cellular membrane is not only a confinement against the harmful outside environment, but it is a thinking organelle. The cellular membrane is ionic in nature, and through the oscillation of its shape, displacement, and charge distribution, it may use this intelligent fluctuation in order to form inner realities by itself, and therefore to hold inner intelligences. And since the cellular membrane separates the inner cell from the outside environment, the intelligences held by the inner realities of cellular membranes specialize in the direct interaction of the entire cell with the outside world, as they are conscious cellular intelligences.

As a reference, you as a conscious intelligence of the human mind are specialized in the interaction of the entire organism with the outside world as it fulfills its needs, or as you fulfill your needs, since it is the same. Yet not only the cellular membrane, but all cellular components, mostly the zillions of proteins and protein arrays, form inner cellular realities, and through them, they hold inner specialized cellular intelligences within inner specialized worlds, specialized in all inner tasks of the cell, counting in zillions. While through these, cells resemble more to living communities and living societies than to individual living beings.

Even more, you can never find individual living beings

anywhere, but only communities, societies, and entire worlds of living beings, all living together in these forms of life, while all forming higher and higher forms of life in this manner. In your own case, you have the following class levels or forms of life: ionic, molecular, cellular, organic, social, up to Life herself. While if you are not aware of these, you may live an entire life associating yourself only with your organism.

At the larger scale of the organism, cellular membranes become skin, and skins become brains. While these components of the physical body have always held the same conscious intelligences for billions of years here on Earth. You have the same specialized intelligences ever since, within these same components of the physical body, performing their specialized tasks endlessly, regardless of the living beings and physical bodies that they inhabit throughout the ages and throughout realities, throughout all generations, species, and forms of life. And I am not considering only types of the intelligences, but the living intelligences themselves, since it seems that it is them the entire time, the same immortal intelligences.

And this is existence, as seen from a physical perspective. Yet it is the same existence as seen from the living and intelligence perspectives, since you can never have one without the other two.

We switch perspectives now to model existence from the Intelligence perspective, since our model seems to be shifting more in this direction. The One has a cognitive nature as we may notice and as all schools of thought claim, yet the One has a living and an objective nature, since you can never separate these three. Yet you will always be tempted to consider the One as being more cognitive than alive and objective, just because you have to employ your own cognitive abilities in order to perceive and understand him. Or this is the case mostly today, since throughout time, people have perceived, understood, explained, and venerated the One through all his supreme perspectives, including his living and objective ones, as Mother Earth, or as his living presence, which is Life herself.

3 INTELLIGENCES, SPECIALIZED EXISTENCE, AND PRIVATE WORLDS

What you want to know now is what these intelligences are. Do they look more as little white snakes, crawling throughout the brain? How exactly do intelligences exist inside the brain? And what are they doing there all day long? And how exactly do you use these creatures throughout cognition, making all these thoughts, feelings and ideas possible?

Intelligences exist not objectively as snakes, but subjectively, more as your characters from the videogames, but without the computer display to see them. While they are still acting there in the mind and within videogames, controlled by you or acting on their own. How do you play your videogame without the display? You cannot, or you still can. You assume and you expect what is going on in the game, then your character lets you know when he is done walking around, fighting, or doing whatever you make him do in there. And then your character tells you what he wants you to do for him and for the game, what he needs, and you do so as best as you can, as you fulfill any need.

Since this is how your intelligences interact with you, by sending you your needs, feelings, and meanings continuously, and so you live your life. Because everything that you do in life you do to fulfill your needs and meanings. Everything. But now, if you do not even know who sends you your needs and why, you might end up neglecting your needs, feelings, and intelligences altogether, and this is not life anymore.

Without your computer display, you cannot know what goes on in the inner computer worlds, since you cannot see what takes place in the inner worlds, yet you may still know, by cooperating closely with your computer characters or intelligences in every manner, by sending and receiving needs and feelings, and by fulfilling these.

And if you do not fulfill each other's needs, you lose the game. Something terrible happens in the game or in this world, you lose your characters and you feel dreadfully, so yes, you find a way, eventually. While you do not even have to know that you have, intelligences or videogame characters, which you must fulfill throughout the game or throughout life, but you simply fulfill your needs exactly as they come, because your intelligences or videogame characters reward you plentifully with good feelings every time you are successful fulfilling your needs. And many times, this is why you live your life, to fulfill these needs and to feel good, or to avoid feeling dreadfully if you do not fulfill them, ignoring the entire time that you play your videogames, that you have a videogame character that you play without even seeing it, or that you even have a computer. Yet you can always manage your existence at all existential levels, since this is the harmonious, intuitive, intelligent existence.

Yet all inner realities are very similar, so yes, this is how you live your life, without a display to see what goes on throughout the inner cognitive realities of your mind. You do your job as best as you can throughout life by fulfilling your needs and by maintaining the living inner harmony alongside your intelligences, in close cooperation with all your intelligences, with or without your awareness. Since all your intelligences interact with you through their needs, which you receive, identify, and fulfill in very large numbers, while assuming throughout life that these are your own needs, and that you fulfill them for yourself.

As stated, intelligences exist subjectively, and many times, they still manage to interact not only within their inner world, but they manage to interact with the outside world, through the matrix holding their inner world. And most importantly, they interact in this manner with you, and through you, they end up interacting with the entire outside world, in every way, for better and for worse.

The matrix of their inner world happens to be the intelligent fluctuation of the electric charge distribution within neurons, and within entire networks of neurons, whenever needed. If you want to understand the inner worlds where intelligences live, you must understand these neurons and networks of neurons first, along with the matrices that they form and maintain throughout life.

All realities are isolated, and while you remain within them, nothing else exists objectively outside these realities, outside your world, regardless of where you are positioned in the wider world. Even within systems or clusters of realities, as it is the case within cognitive systems, realities will remain isolated until two or more realities have to interact in any manner, for various reasons. Usually, realities within cognitive systems interact by the millions, billions, or zillions, when adding all inner realities found within cells, along with all inner realities found within these inner realities, found within subsequent inner realities. And as small as these inner realities may seem from an upper perspective, they are larger on the inside than what you may expect, holding a significant number of specialized intelligences within.

How and why do realities link or interconnect? Realities are specialized, and further on, intelligences within realities are specialized. This is always the case when individual living beings gather to live life together, within cognitive systems, societies, organisms, and within entire civilizations. Their life is easier and safer together, just because everyone is specialized, fulfilling all needs and performing all tasks in a specialized, professional manner, at full efficiency. And this is the case within your organism, with all your cells, since they are specialized. This is the case within the human society, since everybody specializes within society. This is the case within all cells, since all living individual cellular components are specialized, along with all cellular specialized intelligences. Specialized individuals must interact among themselves and with their community, society, organism, in order to gather materials and information, and in order to perform their own task. On a physical level, cells will interact in every manner, sending and receiving materials through the blood stream mostly, and directly through specific tubes, depending on circumstances.

On a subjective level, the intelligences within these cells will interact with intelligences of other cells and realities, by linking their worlds directly. This happens by uniting their matrices. All ions that you find roaming around between cells, through the vibration and fluctuation of their charge distribution everywhere between cells, expand the matrix held by cellular membranes, linking it with matrices of other cells, forming in this manner overall ionic membranes, which are overall matrices held simultaneously by a multitude of cells.

This happens with all specialized cells, including neurons. Neurons are capable to extend themselves physically through axons and dendrites, to reach far away cells and link with them, bypassing unnecessary cells in between. Yet when you study these axons and dendrites, you find them composed of the same cellular membrane, carrying the same matrix present throughout the neuron. And when you study their synapses, you find the same ions put to work in between cells, in order to expand cellular membranes and link their matrices, into a one overall, continuous matrix, forming an overall, larger reality containing all initial realities of all individual neurons. This overall reality allows now all intelligences to roam freely everywhere they need and interact with any intelligence in any manner they choose, while exchanging services, goods, and information, and while enhancing their development. And in this manner, whenever needed, these intelligences form primal intelligences capable to span the entire organism, and even to exit the organism through pheromones in order to perform their own inner tasks in the outside world, or within other organisms, depending on tasks.

These are not cellular networks or neural networks despite of what science may claim, since overall matrices are able to hold and maintain overall realities as a one comprehensive reality.

While there is a difference between networks and overall realities or mainframes, because you always transfer information throughout networks by copying it from one network element to another. But by having one overall reality, you can always move data around freely and accurately within this overall mainframe, without having to copy it. In this manner, you always keep the same original information and the same initial intelligences wherever these need to go. This makes a tremendous difference within Life, since by having a one overall mainframe, by having a one overall reality, you may keep the original intelligences everywhere, without having to copy and project them from one medium to another.

Why do intelligences link their worlds together? They do so only temporarily, since you need to use the abilities of other intelligences found in other realities in order to perform new tasks, and only by linking cellular membranes in very large numbers, you can reach them. Or if you use neurons instead, these are capable to elongate themselves through axons and dendrites in order to reach in a physical manner other neurons containing the necessary intelligence, ability, memory, sense of perception, or understanding found in other neurons or groups of neurons further away within the brain. Because as you notice, not only neurons are used for cognitive tasks, but all specialized cells are, since from a cognitive perspective, all cells perform their specialized tasks in a cognitive, living, intelligent manner.

While if needed, by linking cellular membranes in large numbers, you may exchange intelligences, information, results, ideas, and any kind of goods and materials, all being done on behalf of the entire organism.

And you have to do so, because at the macroscale of the entire organism, the field is very weak, incapable to transfer all knowledge, abilities, or entire intelligences directly through electromagnetic induction, as everything takes place at the cellular level and at the molecular level. But now, by linking in an ionic manner all cellular membranes, you obtain a comprehensive ionic membrane spanning parts of the organism or parts of the brain capable to hold an overall inner reality where all your needed memories, abilities, and intelligences live, and now these know exactly how to organize themselves through their own needs and feelings to perform all unusual tasks of the cell or entire organism.

While you may even see these overall inner realities as they light up on the EEG display, and in this manner, you may figure out what happens in your mind and body. Since from an upper perspective, these are of a subjective or informational nature. While for intelligences, these goods, tasks, abilities, objects, and materials shared within these overall inner realities are objective in nature.

How do intelligences live in there, within their worlds? How do they look like and what exactly do they do there? Inner worlds are exactly as outside worlds, only that they seem different from different perspectives, as all realities do, higher and lower. From upper perspectives, realities seem more as daydreams, videogames, data, information, ideas, reasoning, thoughts, feelings, and memories. Therefore, from an upper perspective, realities seem just as your own mind compared to the outside world. From an upper perspective, you do not actually distinguish individual specialized intelligences at work, doing their duties throughout their life, but you perceive and understand abilities and capabilities that minds have: counting, watching for threats throughout life, reacting to social circumstances, and digesting food. But from an inner perspective, all these abilities are in fact specialized smaller inner intelligences, part of the main, upper, primal intelligence, all living their life and doing their jobs casually within inner worlds held by this primal intelligence, worlds comprising its own cognitive system that this primal intelligence uses for reasoning.

Whenever you see grey 'aliens' performing incisions in the brain with long needles that they insert through the nose or however they insert it, what they actually do, they extract specific cognitive abilities, mostly higher abilities, from wherever these cognitive abilities live within the brain, since these abilities or capabilities are in fact inner intelligences of various kinds and various specializations as seen from an inner perspective, intelligences that might have lived within your genetic line since the beginning, with you never knowing it, and therefore now with you never missing them.

How do intelligences look like from an inner perspective? Intelligences exist in there in an objective manner from their own perspective, yet their existence is always related to their own world. Since all worlds are specialized. Yet since the entire world is meant to be viable, the shape and structure of those inner worlds still makes sense regardless of how these are. And therefore, these inner worlds provide viable living conditions, as different as these may seem from this world. It is similar with all specialized buildings within this world where existence and people in general are specialized, as it happens within pig houses, offshore oilrigs, senior houses, hospitals, or water plants. These are significantly different types of buildings and environments, yet when you are forced to live and work there, even as a novice, you can still find your place and specialization, you can still learn to do your job better, they can still allow you to eat and sleep there wherever people sleep, sit around, eat, and socialize. It might still be too abstract for you to understand living and existence within inner, subjective worlds once you get there, but if you are born there, that is normal for you, while anything taking place here in this world may seem strange, illogic, or entirely impossible. Let us now study these specialized inner realities, meet these specialized intelligences at work and throughout life, and understand them entirely.

First, all these intelligences have a body, since the matrix of their own world along with their continuum provide bodies to all intelligences and objects around there, otherwise, these intelligences would not exist in any form. And now, by having a body, and since this body is objective and physical in nature, since everything is objective and physical in nature within any reality as long as you are there, now, by having both a body and an intelligence, these are genuine living beings, living a normal life within their own world, in their own society, alongside everybody else in there.

Let us study now one of these inner world realities populated by intelligences as though they were genuine living beings in a genuine world. You are already familiar with your inner replica of the world found within your own mind, in the cortex, probably spanning the entire cortex. This is your inner replica of the outside world. First, you as a specialized intelligence are the conscious intelligence as you call yourself, and you are specialized with the interaction of the entire organism with the outside world as it fulfills its needs, or as you fulfill its needs, or as you fulfill your needs, since it is the same. There are zillions of other specialized intelligences in your cognitive system. You are not them and you are not part of their individual life, while you cannot do their specializations either, since you lack the knowledge and abilities.

Your primal eating subconscious intelligence knows continuously what nutrients are within the body, how to manage them continuously, how to get nutrients through digestion, how to manage these, how to transport them to all your cells, trillions of cells, and how to determine you as a conscious intelligence to take the entire physical body and roam around the outside world in order to find more food, more nutrients and calories, exactly as it wants and desires, in that exact amount. Consequently, you cooperate, you feel your hunger or your need to eat, and you find exactly the kind of food that you like to eat then, and you eat it in the exact amount that you desire, exactly as the need states. Doing everything according to your primal eating subconscious intelligence, since it is the one sending you your needs and feelings the entire time. And now you are either rewarded with pleasure and happiness if you are successful finding, cooking, and eating the food, or you are punished dreadfully if you are incapable to identify your need for specific nutrients or if you cannot find them in the outside world being too poor or in a diet. Yet your primal eating intelligence forgives you many times, if you are poor, asking you for alternative food products to find and eat, whatever you manage, containing the same nutrients, and it works.

While you might not even be aware of this entire inner cognitive process of your organism, but you only chase your good feelings throughout life and this is exactly why you always eat, to feel good, or not to feel miserably when you do not eat. Or this is the case when you live your life at the second animal level, because at your third intelligent human level, you already know everything about your mind and body and you know exactly how to find the best food and to eat it at the proper time, only with your main meals, in the most adequate amount. Because at the second intuitive developmental level, you are alive in order to eat. While at the third intelligent developmental level, you eat in order to remain healthy, vibrant, energetic, and fit throughout life, so you may always be able to fulfill your higher needs. Because eating itself is only a second level animal physiological activity.

While as you already notice, you have to eat anyway, so why not doing so at a higher developmental level according to your human nature, while receiving a higher fulfillment just as well. Yet eating is only a second level need, which you may fulfill at the third level or not. Yet you have higher meanings in life besides eating, feeling good, and even staying fit. You have to learn, develop, interconnect positively and prosperously in society, form an entire third level intelligent human society and environment all around, and make this world an equal and more prosperous place. And this is how you fulfill your higher level needs and meanings throughout life and throughout this world.

And in order to do all these tasks, as a conscious intelligence, you have to have an entire inner mind reality only for yourself, where you may learn, remember, reason, mental model, predict, find successful ideas, and implement them in this world in an educated, rational manner. And this is the inner replica of the world, your own, private reality, filled with countless of intelligences necessary to perform your entire conscious learning and reasoning.

While as you notice, your eating primal subconscious intelligence is not in your inner replica of the world, because your inner replica of the world is in the cortex, while your eating primal subconscious intelligence spans the entire organism, as it has its own organs, along with an entire bodily system, the digestive system. And more, because your primal eating subconscious intelligence has to have access to all cells of the organism, since it has to feed them, in the perfect amount, through the entire circulatory system, and it is a tedious specialization. And it can always perform it even flawlessly, with the precision of molecules, because your primal eating subconscious intelligence is a larger system of intelligences, formed of zillions of inner intelligences, all tending to their own inner specialized tasks by the zillions, together forming the primal eating subconscious intelligence itself.

And as you notice, your primal subconscious eating intelligence is not in your cortex where your own intelligent inner replica of the world is, but throughout your entire organism, where it has its own inner replica of its own immediate world, containing your entire kitchen with all nutrients nicely placed throughout cabinets, refrigerator, and throughout the shelves of your supermarket, along with all the procedures necessary to digest them and to determine you the conscious intelligence in the first place to find them, take them home, and eat them, because many times, you do not cooperate, and it is a hassle. Your primal subconscious eating intelligence has in its own inner replica of the world the best knowledge of how to digest these, when to do so, how to take these nutrients to your cells, and then how to further process them there in order to make possible the entire cellular activity possible. Additionally, the primal eating subconscious intelligence knows exactly the amount of food, nutrients, and energy that it needs for all types of conscious activities that you perform in the outside world, since it pays a close attention to everything that you intend to do in the outside world. And if it is more demanding, it determines you to eat more right before you leave the house, or to drink more water.

And if you cooperate and fulfill your needs harmoniously, then yes, you remain successful in everything that you do in the outside world, in the family, at school, at work, or anywhere else in society, since your own intelligences remain capable to provide you with everything that you need according to their own specializations.

While as stated, your primal eating subconscious intelligence can do its entire job flawlessly, with the precision of molecules. Because if it makes mistakes, it leaves the entire organism without food, or with improper nutrients, or filled with expired food, causing all the other intelligences to fail, including you the conscious intelligence.

And as you notice, while you might have assumed that your own specialization as a conscious intelligence was very complex, very important, and very privileged, other primal intelligences have specializations that are just as complex and just as demanding as your own.

Can you do its job in its place as a conscious intelligence? Can you count all these nutrients? Can you feed all your cells directly? Can you take apart every protein of the food that you eat into individual amino acids? Can you produce and store fat? Can you eliminate waste from trillions of cells? No. Your specialization is entirely different, you do not have these abilities, you do not have this specialization, you do not exist as a primal eating intelligence, and what it is more important, you are not that primal eating intelligence.

And this is the case because you are you, you are the

conscious intelligence, and nothing else. You as a conscious intelligence have your own abilities as seen from an upper perspective, while your abilities are also smaller conscious inner intelligences of yours, as your logic, creativity, and continuous awareness.

You are in fact the entire body and the entire cognitive system, with all cells and intelligences, yet you may understand yourself now as a conscious intelligence, to be able to understand your limits, specialization, and capabilities, and you do so by dissociation. Once you are capable to dissociate all intelligences of your cognitive system for who they actually are and not only through what they actually do in the organism, since they are unique, alive, priceless, irreplaceable, and might even have a name. But once you are capable to understand all intelligences of your cognitive system for what they truly are, then you are capable to maintain a continuous harmony within your cognitive system and within your organism. Manage this, and you will live your life problems free endlessly, since you should be able to avoid all problems before they strike you, through a harmonious living. But if you fail to do so, you will drive your body throughout life from ditch to ditch, from illness to illness, from mental breakdown to mental breakdown, and from social problem to social problem, while this is exactly how people live their lives today, everywhere.

Are you more the conscious intelligence, or you are more the physical body? Everything is a matter of your perspective. Besides, all intelligences are conscious of everything, they are conscious of their environment, they are conscious of their specialization, they are conscious and responsible with what they do for other intelligences and for the entire organism, they are conscious of themselves, they are conscious of you, probably more conscious and more responsible than you are throughout your life. Yet if science, biology, and psychology call you the conscious intelligence, since this is how far they can understand the human mind and the human cognition, then we shall refer to you as the conscious intelligence, in order to distinguish you from the rest of intelligences. And this does not stop within the cognitive system, since science and biology call all plants and animals unconscious, for the same reason. While these are conscious just as well. Just try to catch a fly or a rabbit, to see how these outsmart you, being more aware than you are, or simply having more developed abilities to protect themselves.

As a conscious intelligence, you do not live in the outside world, this world where your physical body lives and where your senses of perception are, but since you are an intelligence, you live in an inner world, an inner mind reality, and this inner reality is identical in many ways with the outside world. In fact, you have built this inner world yourself, through your continuous learning and exploring of the outside world throughout life. Because all your memories and knowledge of the outside world in all details are not as books stored on the cognitive shelves of your mind, gathering there the dust of times, but they are arranged in the exact shape of the outside world, resembling the outside world entirely, cognitive brick by cognitive brick.

This does not mean that you have lived your life with a magnifying glass in your hand, continuously trying to perceive, understand, learn, and assimilate everything in this world, in order to form this spectacular intelligent inner replica of the world spanning the cortex, but you did so through your continuous life and experience in the outside world, while learning its most significant details, most significant to you, for various reasons, but mostly in order to be able to fulfill your needs in the outside world. Because everything that you do in life you do in order to fulfill your needs. This is why you have learned everything throughout life, in order to be easier and more possible for you to fulfill your needs the next time you have to do so, through the same procedures, so you do not have to invent the wheel each time. And since all your needs come from your primal intelligences, as your eating intelligence, recovery intelligence, reproduction intelligence, social intelligence, and security intelligence, you are in this together.

You live in your own replica of the world, and you have to understand and therefore interact with the outside world through your own replica of the world. Because the outside world does not even look as you see it now, since everything is grey and blurry. And it is only through your eyes that you are capable to distort the image of the outside world, in order to see it as you do. And you distort it in this manner optically, through lenses and prisms.

Yet if this was your only problem, then you were all right for life. Because there are more problems with the outside world than you may expect. Because this world is so complex and so tedious to comprehend out there, and because that is an entirely different world from your own inner reality, and it is impossible to transfer genuine information from the outside world to your inner replica of the world. Therefore, while you are observing, understanding, learning, and elaborating information from the outside world, you are actually assimilating not accurate information, but only a personalized version of it. And this is more or less accurate, depending on your capabilities, time, interest, resources, constraints, and talent. Yet right now, about ninety-nine percent of your inner replica of the world remains inaccurate, through all consensual beliefs, stereotypes, misunderstanding, strong personal convictions, ideologies, and entire jurisdictions. Because if science did its job right, your inner replica of the world was significantly more accurate. And if society did its job right, now you were more determined to learn and develop throughout life. And if all your groups, entourage, friends, and partners were more careful, now you could live your life drugs free, if only free of pills, coffee, coke, beer, and the rest of the drugs.

Consequently, you never perceive and understand the outside world exactly as it is, but you understand it through your own inner replica of the world and through your own reasoning. And this gives you errors, discrepancies, and disagreements throughout life, since you and all those around see and understand this world in a different manner. They see and understand you in a different manner, you see and understand them in a different manner, while you see and understand yourselves in a different manner. Because you interact with the outside world and you interact with each other through the outside world through your own replicas of this world. All these are different realities. You can never transfer genuine information between them but only personalized copies of information, while they do not resemble. While the difference between the outside world and your own replica of the outside world adds to the erroneous and to the inexistent studied here.

As a reference, take a few minutes to draw your surroundings, and the difference between your drawing and the outside world tells you how accurate your own inner replica of the world is. How important is to learn and develop continuously throughout life? Very. But even more, how important is to learn valid, genuine information throughout life? Highly. How important is to keep out stereotypes and beliefs, and rely only on accurate facts throughout your learning, reasoning, and development? Highly. Similarly, it is highly important to realize that you, as you see and understand yourself to be, are not the individual from the outside world, but you are its replica from your inner world. You are therefore less accurate than who you are in the outside world, and this difference in your own understanding of who you really are decides your entire social life. Because you may live your entire life assuming that you are a normal living being, living a normal life, while in the outside world, you may be something else.

There is still a difference between the genuine, accurate you from the outside world, the image of you as you understand yourself and as you are part of the inner world, and the you as a conscious intelligence performing all reasoning, learning, and understanding of this world. And since you live your life unaware of the three you: real, replica, and conscious intelligence, now these three you interconnect into one, to form you. In other words, you are successful to interconnect three separate existences into one, by linking them tightly unknowingly, and everybody does the same.

This is good and bad, but most importantly, this defines your life now, and it is not exactly too significant. How do you do so? You as a conscious intelligence assume that you are you, the replica of yourself from the outside world, and therefore you adopt all characteristics and abilities of this replica of yourself from the outside world to your conscious being and to your conscious abilities, and you live your life in this manner, as one. You give life to yourself in this manner, since in order to have life, you need an intelligence and a body to contain it.

Is this good or bad? As stated, it is not too significant, but it is only a matter of existence. Because as a conscious intelligence, you are a primal intelligence yourself, and you have been with this cognitive system along with everybody else since protozoa and long before, when you were managing your transfer of nutrients and waste throughout your cellular membrane of your unicellular organism. Throughout the ages, cellular membranes became skin, skin became brain, and here you are now as you read this book, transposed from the outskirts of your cell to the center stage of your brain, in your intact form as you were billions of years ago, coordinating now an entire organism. And if you happen to be a CEO, or the president of a developed nation, then you coordinate most of the outside world, which is an achievement. Yet it is still the same old you as you have been throughout billions of divisions and generations, managing the interaction of the body with the outside world while fulfilling needs.

You do the same with your videogame character, since you associate the image from the display of your computer with its own inner replica from your own inner mind world, which is a genuine, living intelligence. Now that character is also alive, a genuine living being within your own inner replica of the world, and it is the same with all characters of your videogame. And this is when everything starts to happen, because all your primal intelligences cannot distinguish between fiction and reality, or between subjective existence and objective existence at your own outer level, since these are highjective for them.

And therefore, now they assume that all characters of your videogame, along with that entire videogame world and everything happening there is real, everything is part of this world, and therefore they push you now to fulfill your needs within that computer world, since all resources and partners are abundant and available there. And you fall in love with those characters because your intelligences send you oxytocin, you find them attractive, and you take it from there.

It is easier for you if there is a real human being behind all characters, as it is the case when you play the game online, since people even get married in this manner. And when you happen to start a relationship with an actor or an actress, it will add to the overall confusion, since you will always love them not only for what they are, but for what all the characters that they have always played are, since your primal intelligences consider those real, part of your current partner.

It is the same in this world, where each living being from the outside world has a counterpart in your own inner world, and this is how you see and understand them. Even more, since your inner replica of the world is an inner reality in itself, which is a genuine reality, everything is objective from an inner perspective, and therefore all representations of living beings from the outside world are genuine inner intelligences in your inner world. In order to have life, you need a body and an intelligence within. With all details that you learn from all the people in the outside world, now the matrix and the continuum of your inner world are capable, through all these details, to offer a genuine body to all these impersonating intelligences. And this is their specialization now, to be the conscious intelligences of all the inner impersonation of all the people from the outside world, all your family members, colleagues, friends, and loved ones, including videogame characters, consensual characters, news presenters, teachers, movie characters, actors, imaginary friends, all live in a genuine manner within your own inner replica of the world. And this is their specialization now, to live normally within your own replica of the world, while mimicking closely the outside world, as accurately as possible.

If you stop seeing and meeting the real people from this world, the inner impersonating intelligences from your inner replica of the world will still interact normally with anyone there, with you never even noticing. And in order to preserve accuracy, these impersonating intelligences will push you continuously to get news from their counterparts from the outside world. It is the same with all facts and events, because you will always be pushed to understand more about everything within your entire inner replica of the world, upgrading, updating, revising, and renovating it continuously in this manner.

You may have noticed throughout life people living intuitively, unconsciously, through their instincts, at the second, animal level. What they do, they fail to identify their own reasoning, or they fail to apply it continuously for various reasons, they look absent and distracted most of the time, since medicine, drugs, vaccines, beliefs, and additives render you in this manner, and you end up with your own replica of the world and with your own impersonating intelligences in there performing all thinking for you by themselves, without your conscious intervention, dictating now your social behavior and your entire lifelong activity through needs and reflexes. You follow these needs, feelings, and reflexes, as you do not even have to reason anymore throughout life, while in this manner, everyone around including your loved ones are not exactly them, the real ones that you perceive, understand, and interact with, but they are entirely their inner representations from your inner world, your impersonating intelligences. I refer to this as normal intuitive thinking, and it is specific and very common among second level intuitive people, which live their lives at the animal level. And when this happens, all their primal intelligences are capable to access directly the outside world through the conscious intelligence, very easily.

Which is still better than living your life at the first consensual level, which is the servitude level, as it happens when you are part of tight hierarchies as within consensual brotherhoods, political parties, ideologies, or armies, where you are forced to think in specific manners, and you have to disregard the inner simulation of this world taking place within your own replica of the world, along with your own reasoning, since now you have to think and behave according to what they want. Which is better than living your life at the zero addicted level, taking your drugs in an unfulfilled world.

And this is how you learn throughout life, and this is how you elaborate, through the inner intelligences of your conscious intelligence, and through their inner intelligences, all the way down to your basic cognitive procedures. As stated, you may live an entire life without being aware of any of these, without being aware of your entire reasoning, just thinking throughout life as it happens, through any intelligence it happens. And if you happen to live your entire life immersed in beliefs at the first consensual developmental level, or intoxicated in drugs at the zero developmental level, your thinking is never of any interest to you, so why bother?

How do you think? You think through these intelligences, or as these intelligences. You conduct your thinking sometimes as a conscious intelligence, whenever cognition becomes more tedious or more complex, yet you may do so alongside your other inner conscious intelligences. How do you do so? You think, feel, and reason within these inner mind worlds. You happen to have all the needed background information there to conduct your reasoning, within these inner worlds, either within your own replica of the world, or within the inner worlds of any of your intelligences, including your primal subconscious intelligences. Because all your intelligences live in similar inner worlds, which are their own inner worlds, while all these worlds are created through learning, experience, and development, at all levels, and by all intelligences.

Yes, but how exactly do you do so? How exactly do you reason? Your entre conscious reasoning takes place within your inner replica of the world, and it always takes place as mental models and mental simulations. Yet this is how you see everything from an above, correspondent perspecive, as thinking, feelings, thoughts, and mental models, because from an inner perspective, at the level of all these inner cognitive worlds and intelligences, everything is objectively real, with all inner intelligences resembling the actual people, objects, subjects, and events of the outside world.

Even more, the correspondence between your inner, cognitive behavior within your inner replica of the world and your outer, real social behavior is so close, that you cannot distinguish the two anymore, as you maintain them superimposed and correspondent continuously. While this cognitive behavior is so significant to you and to all living beings and intelligences of this world and of the wider world, giving the entire supreme correspondent characteristic of Life and of the wider world. Which for you become the Supreme, Natural, and Spiritual Law of Correspondence.

It is easier to understand your own social reasoning throughout life in this manner, since it is a simple social simulation taking place directly within your inner replica of the world, continuously throughout life. Yet your entire reasoning is a simulation within your replica of the world, taking place throughout its specific segments and domains, as you need it throughout your reasoning.

What happens next at school and at the office? What happens next in your life? You will always know it ahead of time just by reasoning, just by mental modeling your required future circumstances or predictions, within your own inner replica of the world. Is this guy called Ken suddenly appearing in your own inner replica of the world for no important reason? Do your own inner intelligences start interacting with him in every manner within your inner replica of the world, making you jealous? Has it happened before with Joe and Bart in a similar manner? Well, you might be right in your predictions now, because your reasoning never failed while predicting the behavior of your spouse around these guys. And if you are aware of your replica of the world and of all intelligences living there, you may simply watch them interacting while they simulate for you the entire picture as it happens today while you are still at the office, or as it will happen there tomorrow, next weekend, and next month. And this takes place within your own replica of the world whether you want it or not, since your intelligences exist and interact there independently from you the conscious intelligence, ignoring you the entire time. And this is good soap opera, not whatever you see on TV.

Since this is why you cannot sleep at night sometimes, because your intelligences send you your needs now according not to what you actually need, but according to their needs, which manifest as results of these specific inner simulations and inner mental models that they perform there continuously now as they live their normal lives, since it becomes highly relevant for them to live in this manner, night and day. And this is what you have to solve now, even before you can fall asleep. If you take some pills or drink a glass of something, you end up killing these intelligences in mass, along with their entire inner worlds, since these are very fragile and very hard to create, mostly in such accurate details.

But you never care, since no one does, and you just take some pills and go back to sleep. While your problems accumulate, many times exactly as your intelligences predict. But since you lack that computer display to see the entire world clearly, inner, social, and outer, you never identify and realize the complexity of your cognitive system, the complexity of your life, and the complexity of the outside world, to behave accordingly. Yet how could you, since it would mean that you live your life at the third intelligent human level while addicted, while ignorant, while under oath, and while indoctrinated. Which is impossible, since all these are of a different level, and you cannot have your clear intelligent picture through them. But since it is safer in servitude, and since it feels better addicted, this is how you always live your life, at lower developmental levels. And so you end up destroying everything, entire worlds, entire clusters of genuine worlds, destroying your only chance for a genuine reasoning ability at an intelligent human level.

Which is not the case with you, because if you have made it so far in the book, you have managed to understand everything at the third, intelligent human level, since the book kept you within third, intelligent modes of cognition, offering you an intelligent human perspective of the outside world the entire time. But wait until you finish the book, to see life below the third intelligent level, since the outside world itself is underdeveloped, dragging you down with it. And if you still do not believe it, just wait until you finish the book to see it for yourself.

And this is exactly how you reason, through mental models. It is not only you performing the reasoning, but your own inner conscious intelligences do, while they are part of your own conscious intelligence, since you are a larger system of intelligences, mostly as a conscious intelligence. Just monitor your thinking throughout the day, to see that you do not place a single step while jogging without making a short simulation first, of where to place your foot. Because if you do not, you stumble on rocks and branches, and you get hurt. You solve your math problems in this manner, you repair your car, you act socially, and you plan your future, as you do everything through mental models first, and then in this world, but only if you are first successful in your mind.

How do intelligences look in your inner worlds? They are just as normal people, or just as normal objects, circumstances and events, or just as normal concepts and conceptions of these, if they happen to live within inner replicas of the outside world. While subconscious intelligences make specialized personalized correspondences of the outside world, or only of your own inner replica of the world, whatever they need throughout their specialization. And now this is where they live.

While subconscious inner specialized replicas of this world are found by the zillions throughout your mind and entire organism, in each cell, organ, and cellular component.

Yet you the conscious intelligence are specialized yourself, in the interaction of the entire organism with the outside world as it fulfills its needs, or as you fulfill your needs, since it is the same. Which means that your own intelligent inner replica of the world is specialized in the fulfillment of all needs in the outside world. Which for you it might seem that this is the actual outside world, which is not, since it is only a specialized version of the outside world, containing everything that you need to fulfill all your needs and meanings in the outside world.

Yet not all specialized inner worlds and realities are replicas of the outside world, as you make yours as a conscious intelligence, because the inner intelligences of all cells of your body might never see the outside world, but only their own relative outside world, there within the cell or within the tissue or organ. Because this is where they perform their specialized tasks, within the cell or as a cell. That is their outside world, and now that is the replica that they have to create, their own inner specialized world.

As you notice, these zillions of specialized inner worlds and realities found throughout your mind and entire organism in all cells, tissue, and organs of your physical body are actually the cellular components, cells, tissue, organs, bodily system, and therefore entire organism or entire physical body, exactly as it is shaped currently in the human form. Because from a cognitive perspective, you may see intelligences and their entire inner specialized worlds and realities as they perform their own specialized tasks by the zillions, while from a physical perspective, you see the actual physical, objective, material cellular components, cells, and entire organism, exactly as it is today through each intelligence and inner specialized world and reality forming it and shaping it from within.

Many specialized intelligences live in highly complex and highly specialized inner worlds, as the intelligences beating within the heart the entire life, since they have to keep the rhythm, beat faster or slower, according to what you do in this world. Yet wherever they exist, these intelligences are always normal living beings, since their own realities provide them with the necessary physical, objective body throughout their life and existence.

It is important to understand that intelligences create these entire inner worlds, yet they create inner, objective worlds from their own perspective, as they create them through learning and through developmental processes, doing so for their smaller inner intelligences, for them to have a world where they may live and interact, in order to help them perform their own mental models throughout reasoning. This inner behavior is perceived as thinking from an upper perspective, and therefore all these inner worlds have a cognitive nature and cognitive meaning as long as they are created naturally by normal, natural intelligences. Your lifetime learning, discovering, studying, and experiencing are meant to form, update, and consolidate your own replica of the world. You are the creator of an entire world, meant for your own inner conscious intelligences to live, inhabit, roam, fulfill their needs, and do as they please there.

Because as already stated, as a conscious intelligence, you are a large system of inner intelligences yourself, with one conscious specialized inner intelligence for each conscious ability that you have as a conscious intelligence, as conscious learning, conscious deduction, conscious enumeration, conscious elaboration, tactics of succeeding socially, strategies to cook food, ways of surviving recessions, methods of doing homework, knowledge of how to drive in high traffic, ways of making your bed, and strategies of training your pets, as these count in zillions. And now, your own living inner conscious intelligences associated to each one of your conscious cognitive abilities live life normally and objectively in your own inner replica of the world, among all the memories and understandings of all people, objects, subjects, circumstances, and events of the outside world, handling them firsthand in a normal, living, objective, material manner, while embodying them altogether and becoming these throughout cognition, forming their own lifelines of existence by the zillions. Which is the case with your subconscious intelligences, by the zillions, because this is the normal human cognition. As all these inner worlds and intelligences coming from you and from all living beings remain direct part of Life, Intelligence, and the wider world, as these form Life, Intelligence, and the wider world, entirely and exactly as they are today. So what can it ever go wrong?

Because this is your own zoo, or your own civilized world, or soap opera, or prison, hospital, military regime, and democracy, since you always have an inner replica of the world and therefore an outside world exactly in your own image, or exactly as you are capable to form them, depending on your talent, characteristics, environmental conditions, and developmental level.

From your own perspective, the specific behavior and social interaction of your inner intelligences are your own mental models and mental simulations throughout life, comprising your genuine reasoning. And all your primal intelligences do the same throughout their life, they conduct their own reasoning within similarly specialized worlds, full of their own inner intelligences. And all these inner intelligences, zillions in number, in order to reason, they create and maintain their own worlds with inner inner intelligences within, all behaving and living normally in there and probably not even being able to tell the comprehensive meaning of their own life. Since time and space are simple continuums specific to all these realities separately. What takes in this world some time, it will be simulated within inner worlds in an instant, and it will be simulated within inner inner worlds in no time.

And so on, this is the case with all realities everywhere throughout the One, on upper and lower levels. While society and our entire world are one of these naturally created realities, where your own social behavior, your own choices in life, your own reasoning and decisions, your own interaction with your loved ones, your own achievements, successes, and disasters throughout life have a cognitive nature as seen from a higher perspective, all helping some higher being up there to decide specific details throughout its higher life.

Since this is the supreme mentalism characteristic of Life,

Intelligence, and the wider world, or the Supreme, Spiritual, and Natural Law of Mentalism as you know it well.

From an existential perspective, subjective existences may be replicas of objective existences, specialized replicas many times, all meant to help higher beings fulfill their needs, which may be of all kind, coming on all levels. While these subjective, specialized existences also fulfill the needs of all inner intelligences, by providing to them a habitat and the possibility to fulfill their own needs throughout their life.

Your questions now may be of religious or spiritual nature, and this is the case because all realities are similar in existence, nature, purpose, life, and conditions. And this includes our world, this includes all inner realities of this world, this includes our higher reality, and this includes Life, Intelligence, and the wider world as a whole. I study religion, spirituality, and the One in other books of this series.

How do intelligences specialize, when it takes you a tremendous effort to get your education and find a job around here, and to get your career started? And then what you have to put up throughout your jobs and careers, only you know. In order to specialize, intelligences reshape themselves and their entire world. It is no different than you, as you specialize yourself in society, as you specialize your lifetime activity, and as you specialize your entire life in this world. Furthermore, humans modify and reshape their natural environment in order to help them fulfill their needs faster, more certainly, and more efficiently, and this changes their world. I study human development and human specialization throughout other books of this series.

In what it concerns intelligences in general, they will always be eager to live, act, reason, and behave in the most original, specialized, unique, personal manner, since this distinguishes them from other intelligences, giving them their life and existence in this world, throughout all inner and outer realities.

Because specialization is not exactly a chore, a burden, or a tedious job, despite of how it feels where you work here in our world, but for any intelligence, it is an opportunity for life, existence, and recognition. Each new task that must be fulfilled within the organism or in the outside world is an opportunity for any newly specialized intelligence to exist and be part of this world, if they are only capable to figure out how to fulfill it.

If not, then they have to try some more, or other intelligences of the field or Universal Mind have to try hard from below, from within the Universal Mind, to manage to fulfill it. Since this is how entire cognitive systems develop while coping with the continuously changing environment, through all these little inner specialized intelligences figuring out how to form their new specializations according to everything needed to be done in the organism and in the outside world by the organism.

Yet there is more to consider, since Life herself takes you out if you remain underdeveloped, unsuccessful, meaningless, and unfulfilled in life and in this world, since death itself is an invention of Life to help her develop and remain successful continuously. Even more, Life takes you out even if you are old, or if you are only ordinarily successful, or if you are only slowly developing, because Life seeks to keep only the best of the best, the most successful and the most developed among all. While this is the supreme characteristic of meaning.

Yet all these are not the case here in our world, because the current society is consensual, and therefore it is kept out of Life and the wider world, in the Consensual Matrix spanning most of the wider world. Which means that your own specialized activity and behavior in society is not natural, not alive, and not real, but only consensual. Keeping you continuously within the consensual existence, and out of the existence of Life. And now, your current social specialization assigned to you consensually is different than your own natural living meaning that you were supposed to have in Life and in this world, and you suffer in pain and misery every time you go to work. The people at work might be nice, making your long hours at work milder, but if you actually wanted to teach biology throughout life, or developing new specialized drones, instead of patching up roads with asphalt, now you are out of Life, because you never fulfill Life throughout your life, and Life takes you out by taking you right out of her own existence, which is just as dying.

But in any society, natural or consensual, it is very useful to take out the garbage, or to clean up oil pipes, or to repair roads, work in agricultural fields all day, or build a zillion cars needed by a zillion people. Because how could all these be possible if everyone in this world wanted to teach biology? Because this is the current social stereotype, that everything going on in the current consensual world should always be the case, including the multitude of tyrants from the upper social layer. But no, because within Life, the third level intelligent living beings live life within third level intelligent societies, part of entire third level intelligent environments, which are different than the current first level consensual society that you form and maintain continuously alongside everybody else by the billions. Because in the third, intelligent human society, you do not patch up the roads, but you use your intelligence to create the specific drones and entire automated equipment making the roads, cleaning the pipes, and working in the agricultural fields. While you actually get to study biology and the entire accurate science throughout the third, intelligent human society, or you get to perform art, raise children, develop the necessary technology making all drones possible, exactly as your own needs and meanings desire. And if you ever want to do anything else, then yes, you work on everything new, travel this world to be with your loved ones within the comprehensive human family spanning this world, where you may work on any new discovery, part of any new task and specialization that must be fulfilled according to the new environmental changes, but only if you can figure out how to fulfill it. Since this is where all your inventions and discoveries are always helpful, in the fulfillment of all new tasks and specializations. And if you fail figuring out how to fulfill them, then there are a multitude of people like you seeking to figure out the same thing. And once capable to figure it out, this becomes their new specialization in an entire intelligent human world, in the benefit of the entire intelligent human world, which is very, very fulfilling.

But no, none of these is possible today, because the current society and entire human world are consensual, of the first developmental level, part of the entire Consensual Matrix spanning most of the wider world. Since everything consensual taking place here remains the case in most of the wider world. Consequently, now your own task or specialization is assigned consensually to you by someone else, by your master, higher brother, superior, or boss, and this is what you have to do.

Patch up roads with asphalt. Because you could have been the one inventing larger flying drones capable to carry people and freight by air anywhere in this world, making all the roads of this world obsolete, and therefore nobody has to patch them. But now, since your own meaning in life and in this world is assigned consensually at the first consensual servitude level, there will always be roads in this world, always having to be patched, and this will always be your meaning in life and in this world, consensual, patching up roads.

With Life herself keeping you out of her existence now, since you cannot fulfill her by inventing these large flying drones, exactly as she always sends you your needs and meanings to do: research these drones. Or research the mobile power source making them possible, since this interests you since childhood, ever since your parents gave you your first flying toy drone for your ninth birthday. And the thought of inventing flying cars and flying trucks never left you alone.

Because Life is persistent with her needs and meanings. While in your case, and for all humans, you always receive your needs and meanings to develop yourself and the entire world matching your own third level intelligent human nature, because this is exactly why Life has brought you here in this world as a human being, in a human world, to be and to have everything at the third intelligent level, with all the necessary flying vehicles and with all the necessary sources of energy included.

But now, since you remain incapable to fulfill all these necessary third level intelligent tasks and specializations in life and in this world, Life takes you out with the first cataclysm, or she takes you out by default, by keeping you in these consensual servitude prison worlds, where you and the entire humankind cannot reach her anymore. But hey, now you have all these tyrants throughout all upper social layers of the current consensual society, so it must be worth it.

Because now, at your current first consensual servitude level, you lack the cognitive means to develop this world. While you fulfill instead the specific tyrant owning you, since only through you, your tyrant takes over one thousand dollars a day in public subsidies for repairing public roads, fattening him up the entire time. But with you now coming up with this entire nonsense about affordable flying cars, making his own roads obsolete, what is this, revolution? Get back to work and patch up the roads exactly as told.

Why are recognition and independence important to intelligences? It seems that molecular intelligences come in batches, all being replicated in large numbers through transcription and translation according to cellular needs, tasks, and chores, since proteins perform these. Yet even molecular intelligences find ways to become unique in order to find their own unique specialization, their own unique meaning in life and in this world, and their own independence form the other intelligences. This is why you always have to digest every protein into amino acids, since you always need new, unique molecules and molecular intelligences within your own cells, and you can never use those coming with your food as they are, because they are not compatible with your cells, since they were never part of your cells.

How do you specialize as an intelligence, as a living being, and as a human being? Similarly, but you always have to follow a specific pattern throughout your specialized development, otherwise, it does not work. You always start by matching your needs, desires, capabilities, possibilities, opportunities, and vacant positions in this world, in society, wherever you are, and whatever the reality level is where you live. Most living beings specialize in giving birth to new living beings, raising and teaching them how to specialize themselves. Others specialize in specific manners to help others fulfill their needs of all kind throughout life, from eating to transportation, shelter, and coping with the environment, security. We find these on all levels, from subcellular levels to higher organisms, higher cognitive systems, and higher societies.

Yet these are normal opportunities of specialization, because there may be new opportunities for specializations that no other intelligence has ever had. And this is always the case because the natural and social environments change continuously, making available new tasks, new jobs, and new specializations, as you have the opportunity now to lead by taking in these new vacancies, these new positions. And once you manage and know how to do so, how to perform the new jobs, then you qualify as a most unique intelligence, a pioneer in your new field. And since the natural environment changes continuously, this may be the case continuously, many times drastically, as it happens throughout crises, cataclysms, and calamities. Through your own successful specialization, now as a random inner intelligence to fill in the new vacant position and therefore be able to cope with the new environment, now you are successful in making the entire cell, the entire organism, or the entire species capable to cope with the new conditions of the environment.

How do you do so? In many ways, yet you always involve reasoning, as this reasoning is always done through simulations, mental models, and successful ideas. In other words, you always perform your mental models throughout your reasoning repeatedly, until you get your successful idea. This is how you defeat the pathogen, by fighting the battle throughout countless of circumstances, section by section, incident by incident, problem by problem, until you get the entire pathogen defeated. And you have to do so in your own mind as an immune system intelligence, throughout your own inner realities, since if you do so directly in this world, you have the chance to fail and even die. And since zillions of intelligences specialized in the same immune job do the same, they perform the same kind of simulations in their own mind, more or less successful than yours, now all your results and successful ideas are considered in a summative manner, and this is commonly called intuitive reasoning.

Did you kill the pathogen successfully within the inner worlds of your own mind? Because all intelligences have their own minds or cognitive systems, and there is where their own inner intelligences live. Now it is time to apply the cure in your objective world, and hopefully it works. Is the pathogen dead? Now you share the cure and you save everybody, all cells, all organisms, all races and species, and you live happily ever after.

Yet you might not even be aware that your own world has one meaning, to find that specific cure, and that you are a simple simulation yourself, part of an entire cluster of realities performing the same task, finding that cure. And now, since you have all survived, those higher beings above can use your cure, your antidote, exactly as it comes. It works for them, and who knows if they are the ultimate world to have encountered that problem with that plague, or if there are other worlds above them to have that problem.

And it is the same with every condition that your environment will ever throw your way, from draughts to solar flares and mass sterility. All problems in this world, all changes in the environment are simple opportunities for you to specialize yourself in a brand new, unique specialization, to live your life from then on in that specific specialized manner. And it makes you happy and fulfilled. Because once you have nothing to do and remain idle and stagnant, Life discards you in a flash, and you are no more, you join the inexistent. This is different from what the theory of evolution states, yet many things in real life are different than what science states.

What we find significant in this specific topic is the fact that your own existence in this world must match a specific existential niche in the natural environment, while you must match a specific niche in society. And then you have to combine these two, or find a way to match and fulfill them simultaneously. Because existence has a structure, being formed of distinct existential niches. Once you manage to fill in these existential niches perfectly through your own activity, behavior, capabilities, development, specialization, lifelong activity, and entire lifeline of causality, then you are compatible with this world, with Life, and therefore you are compatible with your own existence, you follow your meaning in life, you remain alive, and so you subsist. In your case, this might be giving birth and raising children, feeding an entire community, writing books, building drones, healing people, or assuring security.

Is this good or bad? What exactly is the difference between good and bad in this world? Now that you learn to structure and define existence itself, you are able to identify this difference yourself, without the need of your authorities to define and set it for you. Because many times, authorities define the good and the bad against you, since they follow very selfish agendas.

Our model has defined so far the meaning of life in general, with you part of it, meaning which is to cope continuously with the environment. This is not exactly the case, since this is not the main purpose. When you study the environment closely through all its individual conditions, and when you study all existential niches in this world, you find them not natural and not environmental, since they are alive. Everything is alive in this world, and therefore everything has a meaning, a specialization, a purpose, a set of needs, and an existential niche. This is obvious throughout jungles and social environments, where you have to fight and compete continuously with other living beings only to be able to cope and subsist. If not, you downgrade, decay, and even expire.

4 LIFE IN ALL EXISTENTIAL DETAILS

This world is teaming with life, of all forms, at all levels, and throughout all classes and realities. Science still divides this world into the living and the nonliving, and here is where all errors start, because biology considers alive only organic life, and only organisms out of them, not cells within organisms, and not cellular components. Yet pick a rock and you will find it full of life. There are microorganisms everywhere, even inside of the rock. There are crystalline forms of life held by the entire rock as a matrix, capable to form, hold, and maintain inner realities within, and therefore intelligences. These are of a very low resolution, probably living at a very low pace, but they are alive. It may be insignificant for one stone, but if you consider entire mountains, entire rock formations, and entire planets, you find rocks offering a perfect medium for the crystalline form of life to exist.

You cannot find any object void of life, since you have forms of life at the molecular, atomic, nuclear, and subnuclear levels. From a higher perspective, you cannot separate life, intelligence, and the physical bodies holding them, since these are supreme perspectives of the same oneness. From the smallest perspective, it is the continuum assigning a body to all intelligences, since this is how intelligences exist within realities, and these can never be part of the inexistent.

Life exists in many forms, at many levels, and through a multitude of classes. We have already studied several forms of life implicitly throughout this book. Levels of life are a consequence of the intelligent characteristics and behavior throughout life. I present the entire hierarchy of intelligences throughout other books of this series. For what it concerns humans, you live your life at the zero level if you are very sick, disabled, in a coma, or if you are highly addicted. You also live your life at the zero level if you spend your time in front of the TV with a glass in your hand day after day, year after year, since you cease to exist for Life. What do all these have in common? You are of the zero level if your contribution to Life is zero. Because then it makes no difference for Life if you are dead or alive, since you only consume resources. Note the close relationship between levels of intelligence, life, and levels of existence, just because existence alone may define life and intelligence from supreme perspectives. You may consider these as levels of existence only when you also consider forms of life in the hierarchy of existence. And we will cover forms of life shortly.

At the first consensual servitude level, you live your life on behalf of others, as a slave, as a servant, or as a soldier, or as you do throughout hierarchies as a worker or as a brother, or as you do throughout tight ideologies. Because at the first consensual level, others fulfill your needs, or others allow you to fulfill your needs, while the entire work that you do, you do on behalf of others, those owning and controlling you.

The second existential level is the animal level, comprising all animals, and all living beings capable to move around, including most of today's humans. This is the case because once you are capable to move around, you have your own choice in life of what you have to fulfill, how to do so, and in what order. You may do so in two manners. You may fulfill your needs through win-lose interactions among those around, which is the animal level or the law of the jungle, or you may live your life and fulfill your needs through win-win interactions, which is the intelligent human level, and you may do so through your intelligent reasoning. Because it takes intelligent human abilities to be able to fulfill your own needs while being careful that those around do not suffer when you fulfill your needs. Because when you buy from the grocery store food products coming from the other side of this world where people starve, that is not win-win interaction at all, and therefore that is not intelligent human level. Or when you throw away food for any reason, while others starve by the millions anywhere else in this world, that is not win-win interaction either. Because corporations do not eat, while all consensual laws are meant for corporations and corpses, which also do not eat.

The third level is the intelligent human level. Not all humans are genuinely developed humans, because here on Earth, humans are not born at the human level, but at the animal level. It is only through continuous education, learning, and cognitive development that humans become genuinely developed humans throughout life, if they do so, and they do so by developing their human reasoning. If you miss your chance to learn to speak, for various reasons, now you might not be able to learn to speak and understand speech for the rest of your life. And without managing speech and abstract language in general, you are not able to think at an abstract conceptual level, which is, you are not capable to reason in an intelligent manner. Your reasoning will be done through mental models by your inner intelligences, and you will end up behaving throughout life according to what these intelligences tell you to do, through their needs, punishments, and rewards. Which is exactly the way animals think and live.

Yet life at the second animal intuitive level is by far more developed than the zero and first levels, where you mean nothing for Life, since at the first level you live your life on behalf of others or on behalf of simple ideals, while at the zero addicted level you do not live your life at all according to Life, and you are part of the inexistent.

Further up the hierarchy of intelligences, you have very

powerful intelligences that you can still understand and even interact with, at the fourth, fifth, and sixth levels. Then you have free spirits on the seventh level, you have isolated intelligences living life in the Void at the eighth level, you have the strongest intelligences ever at the ninth level as they live their life tightly connected with the One, and on the tenth supreme level, you have the One himself.

Forms of life are different than levels of life, because levels of life refer more to intelligences and life together, while forms of life refer more to physical bodies and life together. While these both determine existential levels directly, as they determine the existential discrimination directly. Organisms may have priority over single cellular organisms throughout life, while societies may have priorities and rights over individuals composing them. And we see this theme throughout movies, where family members sacrifice themselves for the rest of their family to survive, or for the rest of their nation, society, world.

Classes always refer to individual organisms deciding to live life together, as cells do within organisms, or as people do within societies. It is hard to find the first class to start the classification, since one can never tell how far down the class line life expands. Cells and the organisms that they compose are of different classes certainly. Cells may be of the first form of life, organisms of the second form of life, families and genetic lines of the third class, races of the fourth, species of the fifth, societies of the sixth, civilizations of the seventh, followed by galaxies, realities, clusters of realities, whenever these decide to live together, up to the One himself.

However, forms of life do not start with cells as a first class, since cells are nothing but simple communities of individual living beings, zillions of them. Just because cells are very little, and they have a cellular membrane around, it does not define them as supreme individual living beings. When you study proteins and their living, intelligent behavior and specializations within cells, you find them qualifying as living beings, just as amino acids and ions do. And then no one can ever state that there is no life below the ionic class, because no one can ever perceive what happens at atomic and nuclear levels. Therefore, I always keep this classification of classes only relative to the subject in study.

There is a relationship between classes and realities. Entire classes do not necessarily form overall matrices, capable to hold entire realities, since usually, only selective parts of the entire class do so, as we see throughout cognitive systems. Yet we find entire classes holding continuously a one, overall reality, capable to sustain intelligences, yet this overall reality seems passive in rapport to all the other realities held successively by segments of the entire living class, you never see the entire brain lighting up while thinking and implicitly while managing to form and hold a single, overall reality throughout cognition. Yet you may find overall auras in plants, animals, humans, and in all living beings, auras representing an overall reality spanning the entire physical body, capable to hold intelligences while acting as any medium, as mountains and bodies of water, or as entire forests, cities, nations, or even as entire planets.

Because existence comes in many forms, classes, and realities, and at many levels, bringing with it many rights and possibilities, all giving you more opportunities to live life, develop, and fulfill needs, than what society and those controlling society might want you to know and have.

For example, your third level intelligent human rights are in fact your existential rights of the third level. If you happen to live your life at the third intelligent level, and this is exactly why society sabotages you in every manner, only for you to end up living your life on lower developmental levels, through lower statuses. This is why you see everybody taking drugs and living a meaningless life throughout all movies, in order to determine you implicitly to do the same, and this is what you do. And your status drops. This is why you never see characters learning and developing throughout life, so you never reach the third intelligent developmental level yourself. Or when characters manage to get an education, they tend to achieve it directly, through genius, never having to study, since they always have fun in the movie.

Yet these existential rights are respected by all higher beings, and sometimes they may even intervene on your behalf whenever your rights are not respected. Which is the case with fourth level beings, as angels and superheroes. Whatever the case, angels and superheroes never give you your rights, since only privileges are granted, to first level people, by authorities. Do you have authorities? Then you also have privileges and not rights, since once you obey your authorities, you wave your human rights, and you become a first level person.

As a reference, freedom is a right, while liberty is a privilege. Freedom is what you have, through your own human existence defining you. Nobody gives you this right, and nobody takes it away, as long as you exist, just because you may exist by default, without their intervention. And you always have the right to exist, which means that you always have the right to fulfill your needs. Or this is the case until you give up this right yourself, in any manner. Yet once you do so, once you give up your human rights, you may simply regain them just by stating it, to let anyone know. In contrast, privileges are given to you by your authorities, as long as you agree to remain under their servitude. And if you find throughout your nation symbols, laws, and parties relating more to the word 'liberty' than with the word 'freedom,' then know that everybody is of the first level there, they serve that specific authority, and now they demand this specific privilege from them, while always stating and reinforcing implicitly the fact that they are willing to remain in servitude to them.

Because there is a difference between a person and a living being, and you can be only one or the other. And this is how you exist throughout life, through one existence or another, through one status or another, through privileges or through rights, as a person or as a living being, as a corporation or as a living human being.

You are either a person or a living human being, and this is decided by law, through statutes. And if authorities decide one day to change the statute of the word 'person' to mean non-living entity, then you will not be able to use the word person to define yourself as a living human being, because if you call yourself only person or only being, and not living human being, then it means that you wave your rights, you accept to be a corporation, and you drop into slavery to whatever your authority is, to whatever or whoever owns you. Because corporations can be owned, as any commodity. And this is the case not only with the word person but with a multitude of other words, they form the code of law since language itself is encoded through them, the entire trial becomes not a fair judgment but a simple play of words leading to a simple act or mockery, you lose everything and it ruins your life, while people as those putting you through this entire charade never care and do the same to dozens as you that day. And when the next day comes, they wake up to do just the same to many others. People harming people.

This is a play of words, and it takes place everywhere around and not only throughout courts, resulting in you losing your human status and human rights continuously throughout life, and you end up existing for the rest of your life as a person, which is a nonliving entity, or a corporation, which is a nonliving entity meant for commerce and business, as you are defined by your own identity documents. Or you may end up existing as a corpse, as people have been existing legally in the West since the Romans and long before.

Pay attention to how your documents and your authorities define your existence, because they always state that you are not an intelligent human being, that you are not human, that you are not even an animal, most of the time that you are not even a living being, and even more, that you are not even a simple object existing in this world, as a corpse, but you are always a consensual entity, as a license number, title, or character from a book, since this is what corporations are. Yet they are used for business purpose, which allows those on top of society to own you by default, right from the start.

How exactly do you exist in this world? I study this subject in details throughout other books of this series, since it is relevant to understanding your place in society and in this world. Because who you are depends on your authorities for most of your life, or for your entire life if you ignore all these. Check your documents, and if you find your name always written in uppercase letters, that is not you the living human being, but that is the consensual corporation, which is entirely different than who you are as a living human being. Your existence splits in two from there, and you live your life as two separate beings, one consensual and one natural and alive. While society, authority, and everything around want nothing to do with your natural, living existence, just because you have rights as a living human being. While authorities never disrespect your rights, just because you never consider yourself to be a living human being in the first place. And you even state it explicitly every time they make you do so, through all your documents, and through all your verbal statements, encoded or not. This is how your education is in the name of your corporation, your bank account belongs to your corporation, your children belong to your corporation, if they happen to belong to you. While you have married only the corporation of your spouse and you did so only for business purposes. These things are minutely considered by your authority, and you follow straight through and do exactly what they say, so nobody can ever blame anyone with exploitation, harassment, or murder, since these never apply to corpses and corporations, all these being business as usual for corporations. While corpses are always dead, so they cannot die or suffer. Then how can any of these ever be affected of anything going on in this world? And this has been going on for thousands of years in this exact manner.

Why doing everything in this specific manner? Probably because that those on top of society and on top of this world are highly powerful, highly capable, and highly developed, as sixth or seventh level higher beings, against which you have no chance. But no, this is never the case. Even the fourth level higher beings are angels and they always seek to help and protect you in every manner. If their help ever means anything in all these. And the higher you ascend in developmental level, the more benevolent, harmonious, and potent higher beings become. Those controlling you today are not developed at all, as they are not even genuine living human beings, with you overruling them in rights the entire time.

And it is not even them destroying this world, but through these legal minute details, they determine everybody in this world to destroy others and to destroy themselves, in accord with all laws and regulation, higher and lower. Do they need a highly offensive law instated in this world in order to take down and destroy an entire nation? Then they simply instate Martial Law in the specific building where they sign this law, for as long as they have the meeting and sign the law. And here you are, everything is arranged, since throughout Martial Laws, you may kill and destroy at will in order to fulfill your need for security, in a win-lose circumstance, as it happens during wars. Because during wars, you are allowed to kill enemies without ever being taken to court and charged with murder, but only if Martial Law is clearly instated then, and only if both sides are aware of it. Because all laws are simple arrangements or agreements among people, done for any reason, and now the Martial Law simply states that each side may kill the other.

What can you do? Your life is your choice, because your own existence is for you to decide. Even fourth level beings never help you without your specific request. Yet even fourth level beings stopped showing up long ago, because of this entire legal Consensual Matrix instated all around, and who likes bureaucracy? Who would ever bother to come help a consensual corporation? Since these are someone's property, they are not even objective, and many times, they are not even subjective. With the only detail that these consensual corporations may be owned, just because the money necessary to own them are as consensual as they are, along with the certificates stating all ownership details. And so everything fits together, and you can never have one without the other, otherwise it is not legal anymore. Besides, if you ever find anyone telling you line by line what to do and how to do so in order to overthrow your authorities, and in order to instate freedom in this world from then on, then that is actually part of an ideology, and this world is full of ideologies. Since once you do what they say, then you serve them directly, while they use you to overthrow your current authorities in order to place them in power, and they do the same from then on. This is called revolution because this world revolves in this manner, as authorities replace one another, exploiting you and this world as they please.

What can you do? Just learn more about your actual condition in life, in society, and in this world, and adjust it or change it altogether in a favorable manner, to assure your actual fulfillment in life and in this world. What are these human conditions exactly?

The End

This book series continues with the next book, "The Human Condition." Here is a short synopsis:
The human condition is the print or manner in which you and humanity influence those around, the human environment, the entire world, and Life herself. Furthermore, the specific human conditions that you enhance yourself come back to influence you just as well, knowingly or unknowingly. And it is very easy for me to write in these books how wonderful and harmonious you should be in this world while bringing your most favorable contribution to this world, but you always interconnect with this world as best as you can, favorably and unfavorably, just as those around constrain and allow you, only to be able to fulfill your needs. Since as you study history, some people choose to die than to become dreadful conditions in this world, while others never care, and now this is this world. While there are always tyrants throughout the upper layers of society seeking to be the worst conditions possible in this world, only to debilitate this world, rendering it

controllable, discriminatory, and therefore exploitable.

Since just as you remain aware of all environmental conditions influencing your own fulfillment and condition in life, in society, and in this world, you should remain aware of all conditions that you leave behind through your own life and behavior. Because just as life copes with the environment throughout a continuous toil as science describes, this outside environment is alive itself, it is part of life, it is made of life, and it has you in it just as well, along with your loved ones and your entire condition and contribution to this world. But if you give in and see your environment as a continuously challenging harmful condition since this is what science states, then you engage in win-lose interactions with the environment unnecessarily, while the environment is similar to you, formed of life, people, and the entire human society.

And now, if you become a negative condition of this world by engaging in win-lose interactions with this world, knowingly or unknowingly, the entire world has to cope with you, exactly as science states. While as an unfavorable condition in this world, you do not stand a chance, because you are one and they are an entire world, with or against you. Unless you are organized in this world, to have people against people in this world, which is the most dreadful condition in this world.

Because life, this world, and the human society are very complex, swinging continuously with and against you, since your environment is filled up with conditions that are good, bad, favorable, unfavorable, natural, and consensual, as it is significant to identify, predict, and control them.

Throughout this book, we study the human condition along with all environmental conditions influencing the human existence, as the human condition in life and in this world, the human social condition, and the human higher conditions. We also identify all favorable and unfavorable existential elements along with their consequences, for a better fulfillment.

ABOUT THE AUTHOR

Valentin Leonard Matcas, M.Ed., is a researcher, physicist, mathematician, educator, and an author of nonfiction and fiction books, including the entire "Human" book series. Valentin Leonard Matcas wrote the "Human" book series in the following order: "The Human Needs", "The Human Addictions," "The Hierarchy of Needs," "Stay in Shape, Lead a Healthy Life," "The Human Origins," "The Human Society," "The Human Conspiracy," "The Human Mind," "The Human Reality," "Astral Planes and Your Other Realities," "Life," "The Hierarchy of Intelligences," "The Human Intelligences," "The Human Thoughts," "Mental Models and Successful Ideas," "The Human Attitudes," "The Human Stereotypes," "The Human Ideology," "Modes of Life," "The Human Development," "Patterns of Development," "The Human Lifestyle," "Heal Yourself," "The Human Civilization," "The Human Religion and Spirituality," "The Human Rights," "Higher Laws," "Natural Laws of the Universe," "Existence," "The Human Condition", "Lifelines of Causality," "The Human Behavior," "Flat Earth," "The Human Environment," "The Human Meaning," "The Human Reasoning," "The Human Interconnectivity," "The Consensual Matrix," "The Matrix of Life," and "The Human Knowledge."
Valentin Leonard Matcas writes about terrestrial and alien civilizations, about life in the universe, the way it develops and intertwines across galaxies, about powerful beings as they control and reshape the universe, and about normal living human beings from Earth caught in this beautiful, wider, outstanding interconnectivity. Valentin Leonard Matcas creates a living, warmer universe in his books, teaming with life and vibrancy, on all levels of existence. Valentin Leonard Matcas also wrote "The Storyteller" book series, including "The Storyteller," "Starship Colonial," and "Unlimited," and "The Culling" book series, including "The Culling," "The Dream of the Dead," and "The Last Man on Earth."

When he does not work on his books, Valentin Leonard Matcas enjoys researching, hiking, swimming, kayaking, skiing, snowboarding, biking, reading, listening to music, and playing strategy videogames. You may discover all his books, videos, and articles.